D1508636

Fear of Science—Trust in Science

PUBLICATION OF THE SCIENCE CENTER BERLIN
Volume 19

International Institute for Comparative Social Research

Fear of Science
—Trust in Science

Conditions for Change in the Climate of Opinion

Edited by

Andrei S. Markovits
Wesleyan University
Center for European Studies, Harvard University

Karl W. Deutsch
Harvard University
International Institute for Comparative
Social Research, Science Center Berlin

 Oelgeschlager, Gunn & Hain, Publishers, Inc.
Cambridge, Massachusetts

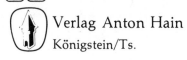 Verlag Anton Hain
Königstein/Ts.

International Standard Book Number: 0-89946-038-0 (USA)
3-445-12053-6 (Germany)

Library of Congress Catalog Card Number: 80-13131

Printed in the United States of America

Library of Congress Cataloging in Publication Data

Main entry under title:

Fear of science, trust in science.

Papers presented at a conference held in West Berlin, 14–17 Sept., 1979, sponsored by the Aspen Institute for Humanistic Studies in Berlin and the International Institute for Comparative Social Research of the Science Center in West Berlin.
Includes index.
1. Science—Social aspects—Congresses. 2. Science—History—
Congresses. I. Markovits, Andrei S. II. Deutsch, Karl Wolfgang,
1912–   III. Aspen Institute Berlin. IV. International Institute for Comparative Social Research.
Q175.4.F42   306'.4   80-13131
ISBN 0-89946-038-0

Contents

List of Tables and Figures

Preface and Acknowledgments

The events of Three Mile Island and Seveso are only recent and particularly drastic manifestations, on both sides of the Atlantic, of technological and scientific failures and their potential for disastrous ramifications. On the whole, however, people of advanced industrial societies have learned to accept the normalcy of automobile accidents, airplane crashes, increased pollution in the environment, and other endemic disadvantages which represent the costs of the very same developments that have improved the conditions of human existence in the course of the twentieth century.

Progress, most certainly, has never been risk-free in any endeavor, let alone in science and its applications to social life. Yet the political expressions of apprehension vis à vis science, leading possibly to its total rejection by certain segments of the public, have assumed various forms of articulation and degrees of intensity throughout history. Some periods witnessed great trust in science and a concomitant faith in technological innovations, whereas others were characterized by fear and pessimism towards the contributions of either. Inter- as well as intracountry differences related to social stratifications, political cultures, and historical experiences have contributed to an additional array of variables influencing the public's attitude toward science and technology. Opinions, in turn, affect the formulation and implementation of policies which, of course, determine not only the future development of science and technology, but also their social applications.

It was this complex set of issues that was discussed with great expertise and enthusiasm by more than thirty international scholars, researchers, and politicans at a conference entitled "Fear of Science versus Trust in Science" held in West Berlin between September 14 and September 17, 1979. Following the festive opening on the evening of September 14, in which the large intellectual framework for the ensuing three days was presented (Introduction), three days were devoted to formal presentations, followed by lengthy discussions concerning the topic's place in the past (Part I), the present (Part II), and the future (Part III). The final session (Conclusion) summarized some of the conference's highlights in a public forum and at a press conference.

This book represents by necessity a fragmentary rendition of an extraordinarily rich conference. Although all eight formal papers have been included, it was clearly impossible to publish the ensuing discussions in their entirety, which meant the unfortunate elimination of many a valuable debate and insightful point. Moreover, the excellent organization of the conference, held in the wonderful setting of the Aspen Institute for Humanistic Studies in Berlin, lent an atmosphere conducive to ample informal discussions which, as most participants assured us, were found extremely stimulating and helpful for their future work. Again, for obvious reasons, these exchanges will unfortunately remain inaccessible to the reader. Thus the discussions following each paper represent *our own* summaries of what *we* believed to be the most salient points concerning each particular presentation and the theme of the conference as a whole. They should in no way be misconstrued as the only valuable contributions to the meetings.

The conference, and hence this volume, would have remained elusive had it not been for the dedicated work and generous assistance on the part of many individuals and some institutions. First and foremost, we owe special thanks to the Aspen Institute for Humanistic Studies in Berlin, which not only cosponsored and organized the entire conference but served as a most gracious host on its beautiful premises. The institute's director, Professor Shepard Stone, ably assisted by James Cooney and the rest of the institute's staff, deserves our gratitude for a flawless organizing effort. In addition to being competent, efficient, and impeccable under the inevitably complicated conditions imposed by the presence of more than thirty natural and social scientists from numerous countries, our colleagues from the Aspen Institute never lost their friendliness, charm, and humor.

We would also like to take this opportunity to express our thanks to the International Institute for Comparative Social Research of the Science

Center Berlin (Wissenschaftszentrum Berlin), which cosponsored the conference. We are especially indebted to Ina Frieser and Konstanza Prinzessin zu Löwenstein, whose organizational talents and indefatigable energy were a decisive factor in the successful preparation and implementation of the conference. Furthermore, we would like to acknowledge the support extended to this endeavor by the city of West Berlin through its Senate, and especially its Senator for Science and Research, Dr. Peter Glotz. Lastly, we are grateful to Irene Kacandes and George Velez of Harvard University for transcribing the discussions, and especially to John Herzfeld, whose editorial assistance proved invaluable to the preparation of the manuscript for publication. Needless to say, they all share only in the positive aspects of this volume. Its shortcomings remain solely our responsibility.

Cambridge, Massachusetts
February 1980

Andrei S. Markovits
Karl W. Deutsch

Foreword

Science and technology in our era have exerted a powerful influence over our lives; in turn, the science policies of governments have influenced the course of science and technology; and, finally, the images and attitudes of diverse publics toward science and technology have influenced the science policies of governments. In this manner, whether we do so consciously or not, we have made science and technology an essential part of all aspects of our daily existence, from government and education to consumption and recreation. Thus it is sensible that we try to find out what we are doing; why we think and feel about science and technology the way we do; what we can learn about this process from past experience; and perhaps what we can do in order to act reasonably in the future.

Toward such ends, an international conference with the theme *Fear of Science versus Trust in Science* met in September 1979 in West Berlin, under the joint auspices of the International Institute for Comparative Social Research of the Science Center Berlin, the Aspen Institute for Humanistic Studies (Berlin), and the Senate of the City of West Berlin.

The conference proved to be a learning experience for many participants, and some of its lessons have broad implications for science policy in more than one country. For these reasons, and in order to invite a broader public to join in the continuing debate and in the policies and actions that may flow from it, the editors have published this volume.

For the Board of Editors of the Science Center Berlin,
Karl W. Deutsch, Director
International Institute for Comparative Social Research
Science Center Berlin

Fear of Science—Trust in Science

The Problem at Hand

Fear and Trust: Contrasting Views of Science in Western History

Karl W. Deutsch

When I was a student in 1930 and 1931 and 1932, I hitchhiked to Berlin, and in those days Czechoslovakia had a government that let me do so. Today the attraction of Berlin has in many ways been more than revived. Berlin is the kind of city of which an American would say, "It is a place that makes you think." We have taken that literally. So has the Aspen Institute, so have the people from many countries who have come to the Wissenschaftszentrum Berlin, the Science Center Berlin , including the young Americans who work there side by side with young German scholars, in the particular group with which I am working in the International Institute for Comparative Social Research.

In a major part of this work we are trying to find out where the world is going, and we have found it a fascinating and difficult subject. We have set up some world models, and one of the problems that came up was the rate of scientific and technological change. We have come to suspect that technological change, to a certain degree, depends on whether people want it to happen or not. To some extent, change occurs even if they do not want it, but public support or public hostility can make a very large difference. Hence, our interest in the topic of this conference.

Note: This chapter represents part of an address delivered by Karl W. Deutsch at the dinner ceremonies opening the conference.

I will offer you here only a brief overview of history, since there are enough eminent historians among us to correct me in the following three days of our meetings. But at least some historians report that about 190 years ago, back in July and early August of 1789, there was a movement called "The Great Fear" among French peasants. Those peasants, it seems, were not quite sure of what they were afraid. Some believed that bandits were coming; others were afraid of wolves. Whatever the local details, they were very much afraid. It later turned out that this Great Fear of the French peasants was the prelude to their getting rid of the *corvée*—the forced labor of serfdom—and that they eventually became free owners of their land and, still later, mainstays of conservatism in French politics. To me, the story of this episode has meant that the Great Fear of 1789 was a prelude to a very great change, and I have never forgotten the seeming historical paradox: that times of great change may have an early stage of great fear.

I am thinking of a second story. I do not recall all the precise details about it, but the story goes that three medieval monks were told that a very great treasure had been buried on a very high mountain peak, well above the timberline. Eagerly, these three friars climbed up and out of the woods, got promptly lost in a dense fog and a very cold storm, and concluded that up there no reasonable creature would ever go, that this was obviously an area of the devil. They came down and warned all the people never to go up there again. About 400 years later, in the eighteenth century, about 1760, when more than one-half of the English population lived in towns, Englishmen began to climb the Alps and eventually the Matterhorn. Since then, climbing the Matterhorn has become a standard, albeit difficult, part of every devout Alpinist's repertoire. So perhaps evil is not quite so present up there.

What a contrast we find when we go on from that medieval fear of nature, that fear of the unknown, that fear of the new, and then come to the new confidence that prevailed from the Renaissance onward and reached its peak in the Age of Enlightenment, when people like Lessing and Kant wrote splendid essays that belong among the great pieces of world literature, picturing how mankind was going to progress from enlightenment to enlightenment, and how science would try to give us truth.

And then we encounter the split personality of the nineteenth century, the century in which William Blake and James Watt were contemporaries. It was the century in which Blake wrote about the dark, satanic mills but engineers built them; the century in which the city of London is pictured in 1820 by the cartoonist George Cruikshank as a huge giant trampling forward into the countryside, crushing down the green fields of nature, hurling showers of bricks. This cartoon, entitled *London Going out of Town—or—The March of Bricks and Mortar,* has been reproduced as the frontispiece in Arnold Toynbee's book *Cities on the Move,* and I could imagine meetings in Germany and in the United States where it would still be considered very timely. This was the split personality of the nineteenth century, where science advanced and romantics deplored it, when the railroad engineers built railroads and for almost every railroad tie there seems to have been at least one Romantic poet or writer extolling the past. Two contradictory sets of thoughts and feelings existed side by side.

Then came the twentieth century with more machines, more technology, dynamos, automobiles, diesel engines, the good ship *Titanic* with its confidence and with the fate in which it ended. (By the way, there is an almost verbatim description of the fate of the *Titanic* written in the early 1840s by the Danish thinker Sören Kierkegaard. The classic symbol of the risks of science and technology was seen by a writer long before the actual event came around.) Then came the First World War, the tremendous wreck of world civilization. Mankind staggered back into some sort of reconstruction, and then dropped into a second catastrophe. Nevertheless, there was once more a revival of a certain belief in science. It began in the 1930s with a science movement, when many scientists, obviously short of funds, discovered how important it is to explain for everybody the significance of science. They became excellent popularizers of science from the thirties onward, and there followed the great hopes of sulfa drugs, penicillin, hybrid maize, nylon, and much else in the 1940s and 1950s. Science for a time once again was associated with human happiness.

What came later was the discovery of nuclear energy and the slow but growing reaction to the use of the atom bomb. It was a change of mood foreshadowed in J. Robert Oppenheimer's thought, "We have known sin." Again it was widely felt that science was terribly dangerous; we are now back again in an age when it is fashionable to condemn it.

Is it this time a complete turning away? Are we trying to go back to the Middle Ages? I know of sensitive and intelligent young people who have tried for a year to live a medieval life. There is a young humanist scholar who is busy running a farm in Maine and killing sheep in order to be "close to nature," but he is paying for the deficit of his farm with earnings from translating literary works.

In the midst of all this, are we going back? Or are we once again, like those French peasants in the age of "The Great Fear," on the threshold of very great changes? It is easy to ask; it is hard to answer. I have only told you briefly *what* happened—the historians present will describe the events much better, and will show how much more complicated it all was than this very simplified overview has described.

But even this brief sketch may lead us to ask *why*: why do people sometimes trust science with enthusiasm, as Benjamin Franklin did, and why are they sometimes deeply afraid of it, as Sören Kierkegaard was? *Why* do we feel one way or the other? When are antiscientific movements a major political force to be taken into account carefully at the eve of a major political decision?

We do not know. We would like to know. We must ask the historians. We must ask the experts of public opinion research, both from Europe and from the United States. We must ask everybody who can give us any piece of the puzzle, and perhaps, just perhaps, with much good luck, there may be a chance for us to end up just a little bit less ignorant than we are now.

Chapter 2

Science, Politics, and Ethics

Peter Glotz

The hope that scientific progress can solve our problems no longer predominates in the political discussions of the Federal Republic; lately it has been the danger emanating from science and technology that has increasingly taken the center of public debate. The top places of the bestseller lists have been filled for several months with books insisting that only a new way of looking at life and the world will free us from the paradox of our time: that man, in order to survive, is forced to use up more of his resources day by day to protect himself from the repercussions of progress: that is, from himself. There is a danger that fear of science, fed anew by each environmental catastrophe, may lead us into a flight to totalitarian ways of life; already there exists the dark fantasy of an ecological dictatorship. Therefore I am particularly grateful that the Aspen Institute has chosen this topic for its conference.

In 1479, a proposal was put forth that the government of Danzig should drown the inventor of a loom that could weave several pieces of cloth at the same time; in 1978 and 1979, protest demonstrations against nuclear power plants in France and the Federal Republic led to conflicts resembling civil war. A few weeks ago, poisonous fumes from a cement factory polluted a whole district in the Federal Republic; soon, with the

Note: This chapter represents part of an address delivered by Peter Glotz at the dinner ceremonies opening the conference.

aid of genetic technology, scientists will probably be capable of using bacteria to produce a drug to counteract cancer, a disease that can be caused by pollutants in the environment and which occurs, to a frightening extent, more and more frequently among children. There are 500 years of scientific development between these randomly chosen examples; science today is a determining factor in our epoch, and it has almost become the characteristic of our age.

The inventor of the loom was not drowned. In the fifteenth and sixteenth centuries, the value to humanity of new discoveries was only hesitatingly recognized. Belief in progress came very slowly; the mentality of those times did not allow for the promise that general prosperity and happiness could occur on earth. Before the European sensibility came to accept the mythos of an ever-happier future on earth, it passed haltingly through many stages; the spiritual basis of these stages was always the concept of happiness assured to them only in another world.

The initial skepticism, the otherworldliness of thought, above all the certainty of a day of judgment which could come at any time, first faded after Descartes; it then quickly gave way to the realization that, with the help of science, yesterday's incredible dreams could become reality. In deference to the persistent threat of hellfire held out by the Church—Galileo's works remained on the *Index Librorum Prohibitorum* until the beginning of the nineteenth century, and he has still not been rehabilitated—the price science had to pay was political abstinence on social questions. Only in that way could it escape persecution. The natural sciences thus retreated to more sophisticated experiments whose focus became ever more exact; this sophistication fostered their triumphant progress into the nineteenth century. The humanities followed the ideal of the exact natural sciences.

The natural sciences helped more to drive away the belief in devils than did religions and philosophical enlightenment; they explained to mankind, with increasing plausibility, the world and man's position in it. The demystification of the world had such an effect—for example, in the field of anthropology—that traditional concepts like "good" and "evil" were explained and replaced by concepts like mental illness, frustration, inhibition, and aggression.

Science resisted the pull from its apolitical attitude of nonevaluation even after the threat of immolation at the stake disappeared. Such resistance finds its clear expression in the writings of Max Weber, who assumed that science as such could not determine values. I feel very close to Weber's skepticism. However, I also believe that this restriction of scope led to a basic refusal to further discuss theoretical applications, which spread like wildfire to become a resigned and tacit acceptance of the demonic nature of questions not resolvable in favor of "final purposes." The psychological effect of Weber's theory on Europe is dangerous: we cease our discussions too soon.

Belief in God, in his unusual and marvelous ways, gave mankind a feeling of total security for centuries. A believer could attribute catastrophes to a higher order of things, which he could not understand but in which he knew himself to be included by God's almighty goodness and reason. Religion could not remove his suffering, it could not prevent his death, but it could make them acceptable. Modern science has contributed to the dissolution of this assurance of identity; it has shown that norms have become directly manipulable. Moreover, it has banned norms from scientific thought entirely, unless they can be proven on a rational basis. Science has handed over to politicians the task of applying its findings to society; in return, politicians have agreed to guarantee an almost unrestricted policy of *laissez-innover*. Ceremonial addresses at the annual celebrations of great scientific organizations bear continual witness to this division of responsibility.

Today, as technological supersystems demonstrate their power, they produce a consciousness of helplessness and passivity in citizens. Environmental catastrophes can no longer be ascribed to God's unfathomable ways. To the contrary, they are directly attributable to science. After a period of unlimited confidence in science, people have begun to resist it. Opponents of science, whose conception of themselves remains rooted in the nineteenth century, are aligning.

I want now to depart from these rather extremely formulated historical restrospections; I am no historian, but the problems of the present, the uneasiness with science, which is more and more strongly articulated politically, can be neither explained nor resolved without an appreciation of history.

Whereas people of the fifteenth century could understand a mechanical loom, people today feel powerlessness when they are confronted by modern technology, because it is no longer commonly comprehensible. It fills them with fear. At the same time, their trust in those who do understand and control it is vanishing because people see themselves as standing at the mercy of scientists. This is often attributed to the assumption that the development of man's personality has lingered behind technological development; this argument says that the nature of man's personality should be altered to conform to today's standard of technology. That this is possible, some say, has been shown by the example of the computer-controlled astronauts, whose bodies and minds were conditioned to function in space. But this conditioning of human beings is far from educationally ideal. It can be universally achieved only by an authoritarian training which surpasses anything previously imagined. Man's ability to adapt should not be overestimated; this is illustrated by the withdrawal from society of groups in the population who resist the changing norms of everyday life by maintaining an aggressive distance from it.

The possibilities unleashed by genetic technology are no longer visions of horror but will perhaps soon be reality: reports have already come from America showing that it can be established from human genetic structure whether or not a subject is well suited for a particular kind of activity. The idea of genetic stigmatization, of the upgrading and downgrading of people, does not exactly encourage confidence in science or in its capacity to cope with that progress which it itself has created. In this respect, I have only limited trust in the thesis that absolves the individual scientist of responsibility for his findings, when it is only his civic courage that determines whether or not he makes his findings public. It was in the recent discussion of the legal regulation of genetic research that I became convinced that scientists, allied to industry as they are by strong commercial interests, are sometimes prevented from reflecting on the danger of their own research and also from renouncing that which is *possible*, but not desirable.

Viewing such realities, people come to think that, when faced with a technology that develops, so to say, "naturally," politicians and perhaps even democratic institutions themselves are not able to solve the problems of scientific progress: energy crises, environmental catastrophes, terrorist attacks against or negligent operation of large nuclear plants. It is only a short step from this perception of helplessness to the demand for an authoritarian state which makes all the decisions and distributions and which substitutes obedience for identification with the state.

I consider cynical the silent hope of many people that good can only be expected to come from evil. Anyone who longs for another Three Mile Island, anyone who hopes for a devastating epidemic to result from an accident in a genetic engineering laboratory, so that society will be forced to change its views because of the catastrophe, is just as unconvincing as the person who blindly trusts science to overcome all public crises and withdraws into his private life.

Many justified hopes are set on a new scientific rationality, on a better education for mankind brought about by a socioeconomic research program that would make it possible to overcome the problems associated with scientific progress. But rationality is not a way of seeing the world, not a substitute for ethics. It is also a mistake to think that increased irrationality, a refuge into exaggerated sensuality, is the answer to a mechanical, rational technology. A person who sees no possibility of changing something seeks escape in a moodily savored depression and continuous reflection. Youth sects, seeking refuge in subcultures and new religions—but also the "great refusal" of many intellectual circles—are reactions to an unmastered scientific and technical development.

We must reach a new view of science, and the sciences must come to a new view of themselves. Science can no longer deny that it has practical and political, as well as ethical, consequences. The scientists, because of their technical knowledge, must be the first to give warning, since they can anticipate the problems that result from a successful application of their work. The values of a society, the new and necessary scientifically based ethics, must not be outside science but must be a part of it, its very orientation.

I am optimistic: we do not need to drown our scientists. The promise of public self-determination exists. The renunciation by the United States of the development and construction of a supersonic passenger plane is to me an outstanding decision of recent years. A democracy highly developed in its technology was capable of renouncing something both technically possible and industrially desirable. This renunciation offers hope for new variations on the SALT talks, providing for limitation not only of armaments, but also of industrial, technological, and scientific developments as well. Industrial and technological "arms limitation" is not enough; in many areas, complete "disarmament" must be achieved. We cannot abolish scientific progress. And we ought not *want* to abolish it. However, we must cease the race that may soon bring us into crisis.

Though I am not without hope, I will not deny that I am skeptical. Think of the situation in the Federal Republic: we want to curb our energy use—in order to avert deterioration into a total surveillance state —and we want to keep the option for nuclear power open but to limit considerably the number of nuclear power plants. This requires a new energy policy, one which, by the way, must withstand the strong opposition of the energy industry. We want to limit emissions in the air; above all we want to live in a clean environment. However, this means, among many other things, that we must impose numerous conditions on industrial production, and that we must probably restrict the use of automobiles. Mentioning only these two problems, the question must be asked: will we obtain the consent of the majority for the necessary measures? How will these measures harmonize with the demand we all make for full employment, and with the unwillingness on the part of the overwhelming majority of our population to accept reductions in our standard of living?

These are, moreover, only two sectors. Let us take communications technology as the third. I am completely unconvinced that we would do people a service by offering a choice of thirty television programs in the future instead of three or five; therefore, I believe that the application of some new electronic techniques should be blocked. At the same time, I ask myself: if we stop nuclear energy, if we are to take new approaches to traffic policy and therefore—at least in the medium term—limit the automobile industry, can we then also place restrictions on communications technology?

I am not claiming that these problems are insoluble, but their solution requires that scientists be willing to join in dialogue with those who are afraid, and that those who are afraid be willing to join in dialogue with the scientists. These two conditions are seldom met together. At a recent general meeting of a large German scientific society, I heard a scientist, angered by criticism from the ecology movement, simply lump the so-called green people together with Nazis and Communists; he literally spoke of the "brown, red, and green," as if they were all the same. Does such a scientist think he is contributing to the solution of society's problems by making such pronouncements? And there are, of course, such people on the other side: people who shut their ears, who no longer believe in God but nonetheless make nuclear energy out to be the devil, and who are no longer agreeable to impartial discussion. I hope that this symposium will contribute to making such attitudes as rare as possible and that a dialogue between scientists and the public will soon be obtainable.

In my private opinion, a new ethics based on science is necessary for this to take place. If we are not prepared to attempt to argue over our "final purposes" in a way that can achieve consensus, then we will fail. Here I see a new role for philosophy: if it is not able to play the part of an interpreter or moderator between one expert whose horizons are bound by his subject, and the next, then we will continue to talk at cross-purposes.

I want to stop here, as it were, in the middle of a sentence. The scientific optimism of the nineteenth century is gone; I advocate doing all we can to avoid sinking into cultural pessimism now. We stand together —scientists and politicians—confronted by a very difficult business. I hope that this symposium will make this business a bit easier for us all.

PART I

Historical Aspects

Wisdom, Science, and Mechanics: The Three Tiers of Medieval Knowledge and the Forbidden Fourth

Robert S. Lopez

Not a scientist or a science historian, not a political scientist or a philosopher, but merely a general medievalist interested in the infinite variety of human thought and behavior, I accepted the invitation to this conference with warm and rash enthusiasm. The theme was imaginative and challenging, as is everything conceived by Karl Deutsch, and I was looking forward to its discussion. But as I proceeded to prepare my report, I began doubting the wisdom of my decision, noticing the gaps in my science, and wondering about the mechanics of an inevitably sketchy presentation of a facet of a theme to eminent specialists of entirely different facets. I cannot resort to the forbidden fourth tier of medieval knowledge—magic—and it is too late for me to withdraw, but there is time enough to invoke the mercy of the jury.

Were we to deal with the historical dimensions of the love-fear complex about science merely to provide practical solutions for the future, it would be proper to adopt the modern definitions of pure and applied science. On these terms of reference, however, the answer would be unhelpfully brief. We are not afraid of pure science, which we identify with the dispassionate quest for truth, but uneasy about applied science, or technology, which is powerful but may be used for evil as well as for good. In the Middle Ages, the opposite view prevailed: there was fear of scientific speculation for its own sake, a legitimate but dangerous path which could lead closer to God, the supreme truth, or mislead to the Devil; they were not afraid of applied science or technology per se, which was weak and vulgar but hardly a threat to salvation.

To the best of my recollection, only one device of medieval technology was condemned by a council, in 1179: the crossbow was deemed an inhuman weapon, no doubt because it was then used most effectively by townspeople to pierce the armor which had long made their ironclad lords unbeatable. Naturally the burghers refused to play war according to rules not of their own making; soon after, the Teuton Knights found the crossbow handy to crush the resistance of the Estonians, who anyway were pagans; the kings of Aragon employed crack crossbowmen indiscriminately against Muslims and Christians. Indeed the crossbow was a fairly sophisticated machine, slower than the longbow but far more precise and powerful. If the crossbowmen in the French army at Crécy were overwhelmed by the English longbowmen, it was because they were too few and not properly deployed, like the tanks of Gamelin in World War II.

I have lingered on this isolated example because it seems to offer a certain analogy to the problem of nuclear power—is it at all possible to stop forever the propagation of an inhuman but effective weapon?—but the resemblance is superficial and does not really answer the problems of our future. At the very worst, the crossbow could have undermined the power and prestige of the privileged classes, but not ignited an apocalypse. We cannot fathom the historical dimensions of the medieval ambivalence towards science and learning unless we adopt its definitions. These are not entirely superseded: the official doctrine of the Catholic Church still clings to them.

In the latest edition of the *Enciclopedia Cattolica*, published in Rome by the Jesuits, one finds a description of three attitudes towards the quest of knowledge, which is lifted bodily from Aquinas with a few complementary flashbacks to St. Jerome. Too little interest in knowledge is "culpable ignorance," a vice; a natural amount of interest is "studiousness," a virtue; an excessive craving for knowledge is "curiosity," a sin. How can one be sinfully curious? Apart from the trivial trespassings of the flesh or the mind—which bear no relation to what we now call science—sinful curiosity occurs chiefly in three ways: if one seeks knowledge for pride or with an evil purpose; if one does so with forbidden methods such as necromancy and divination; or, above all, if one pursues knowledge in order to pry into God's secrets, namely, the mysteries of the faith, the end of the world, and the hidden intentions of the Lord.

Before elaborating on these warnings, which would require some interpretive effort to be made comparable to the modern climate of opinion about scientific pursuits, let us go further back to the medieval roots, from Aquinas and the thirteenth century to St. Augustine and the fifth century, when the Christian definitions of wisdom, science, mechanics, and magic were reshaped through a cautious reinsertion of classic philosophical concepts into biblical teachings. Then, as later, the word *scientia* (from *scire*, to know) meant knowledge of any kind, whereas *sapientia* (from *sapere*, to taste and hence to be wise) meant wisdom in a moral rather than intellectual sense. But the two words were also used interchangeably and had been tied in an immensely popular Stoic formula (so popular that jurists had borrowed it to describe their own idealized science): "Wisdom [or jurisprudence] is the knowledge of human and divine things."

Augustine accepted the formula but split it: the proper definition, he wrote, is that sapience is the understanding of divine, eternal things; science is the knowledge of human, temporal things. Thus secular learning is definitely downgraded—*scientia* below *sapientia*, which is not surprising in the context of that writer—but it preserves a dignity of its own. Augustine knew from Plato that man's innate ideas, derived from secular experience, ultimately led to his divine models, and knew from Aristotle that man's instinctive urge to learn about things through their causes finally led to God, the cause of all causes. Nevertheless, he stressed that it was not essential "to know the causes of the world's great physical phenomena hidden away in the most secret recesses of nature," and that the "easily readable" language of the Holy Scripture enabled even the babes to "rise gradually to divine and sublime matters" without any need of secular teaching. Implicit in these statements was a profound diffidence towards human nature and judgment. At the basis of man's knowledge of good and evil there was the original sin, the fruit of the forbidden tree; for the sake of that fruit man had connived with the Devil, tried to become a god, and condemned Christ to the agony of the Cross. If the quest for rational knowledge had led mankind so far astray, science was to be feared, sapience sufficed.

As for applied science, the only form of it that medieval opinion (and not only medieval: the belief continued well into the eighteenth century) credited with the capability of lending man a significant control over nature, was magic. Virtually nobody dared defend it—it took a freshly converted Spanish Jew, Pedro Alfonso, to mention, in the early twelfth century, prognostication and necromancy as matters that some unnamed philosophers would regard as the seventh liberal art—but hardly anyone questioned its power. It was a damnable thing, to be sure, and yet the sacred rites of Christianity also involved appeals to the supernatural, so much so that St. Martin of Braga, in the sixth century, endeavored to wean the Portuguese peasants from the worship of devils by encouraging them to make the magic sign of the Cross and the "sacred incantation" of the *Credo*. Between a mass and an exorcism, and in spite of unqualified condemnations by the theologians, many priests indulged in minor magic practices, which would now appear both ineffective and harmless. Many pursuers of learning, possibly most of them, included forbidden "curiosities" in their research; so long as they did not sign with their blood a contract with a devil, the sin was not unredeemable, and a discreet sinner was seldom prosecuted.

How about mechanics, or, as one would call it today, technology? It was morally unobjectionable, immensely useful in its practical achievements, but, in most of its manifestations and for most of the Middle Ages, weighed down by the fact that it entailed manual work. The prejudice went back to classic antiquity—characteristically, Seneca confessed to admiring sculpture but despising sculptors because they dirtied their hands—but it deepened ever since Cassiodorus, in the sixth century, eliminated from the gentlemanly course of studies the only practical crafts that Varro had included in it, medicine and architecture. Six hundred years later, when Hugh of St. Victor tried to put in a good word for what he called the "adulterine" science of mechanics, he recommended it only as a plebeian study, intended to cater for material needs. Obscurity had its advantages: unwatched by thinkers, the tinkers had remained comparatively free to ply their trades according to their wits, and, through manual and mental cleverness, to produce the new or improved tools and techniques that enabled medieval Europe to soar higher and to thrive far more than had the ancient world. So long as there was no dialogue between engineering and philosophy, however, no practical invention led to those dangerous but exhilarating breakthroughs that only pure thought can perceive and exploit.

There is no denying that, throughout the Middle Ages, mystics and the sterner seekers of sapience endeavored to throw secular learning on the defensive and the mechanic crafts into the shade. St. Peter Damian upbraided scholars who studied letters to improve their understanding of God, as if one had to light a lamp in order to look at the sun; St. Bernard of Clairvaux denounced the exuberance of church decorations that distracted the monks from prayer; St. Francis of Assisi admonished, "Even supposing that you have sufficient science to know everything, every language, the course of the stars and all else . . . one single demon knows more about them than all mortal men together; yet there is one thing of which the Devil is incapable and which constitutes the glory of man, loyalty to God." Indeed the difference between "studiousness" and "curiosity" was not easily perceived: a thousand years ago, Gerbert of Aurillac (Pope Sylvester II) was one of the earliest and most prominent scholars whom posterity suspected of magic practices because he knew too much for his time, including bits of Greek and Arab science.

Yet the medieval Church, perhaps through inefficiency more than design, also had surprising areas of toleration. In a striking pronouncement that Galileo was to quote in his defense, unfortunately to no avail, Augustine himself had suggested that "if we happen across a passage in Holy Scripture that lends itself to various interpretations, we must not . . . bind ourselves so firmly to any of them that if one day the truth is more thoroughly investigated, our interpretation may collapse, and we with it." There are paintings that show him or Aquinas trampling upon Aristotle or Averroes, but most philosophers and scientists made reference to opinions of their pagan, Muslim, and Jewish counterparts more frequently and safely than Americans could quote Marx before certain committees of the McCarthy years. What is more, in the thirteenth century, Bishop Grosseteste and the other philosophers of light explored successfully the physical causes of the rainbow, with no reference whatever to the assertion in Genesis that it was God's own bow, placed in the clouds by him as a self-reminder that he had promised Noah never again to unleash a Flood.

The truth of the matter is that most medieval thinkers accepted the limits set by God to their quest for knowledge, but within those limits handled their God-given investigative skills with growing confidence. They did not need to grope, like Einstein, for a unified theory of physics; God, they felt, provided that unity. He was the order of the world, the harmonizer of all contrasts, the initiator of all motions. Love gushing from Him to all creatures and returned to Him by the creatures in proportion to their abilities was (as one would say today) the basic source of energy. Made in God's image and placed at the center of the Creation, man was indeed an imperfect masterpiece—in Augustine's and Dante's metaphors, a worm directed to becoming an angelic butterfly—but he was capable of discerning the purposes of his Creator by studying their visible imprints on the Creation. As long as he shunned the diabolic temptation of trying to learn too much too soon, he should not fear science, but trust it and love it.

In the early Middle Ages, God's purposes were hazily perceived, mostly through far-fetched allegories; thus, in the sixth century, Gregory of Tours interpreted the falling of leaves in winter and their return in spring as symbols of Christ's death and resurrection. Allegory never lost its popularity: it was a favorite literary device, and its very vagueness came in handy to express ill-defined or dubiously orthodox concepts. The great revival of the late Middle Ages, however, brought out a fresh wave of optimism and a strong yearning for more precise, concrete, and rational (though not rationalistic) exploration of the natural world. Faintly audible as early as the tenth and eleventh centuries, when the practical sciences of medicine and law regained a place among humanistic studies (indeed, above the basic curriculum and only slightly below theology, the third "graduate faculty"), the protest against passive acceptance of revelation and authority as the only sources of knowledge became very loud in the twelfth century. "I take nothing from God, for whatever exists is from Him and because of Him, but the natural order does not exist confusedly and without rational arrangement: human reason must be listened to," wrote Abelard of Bath. More crisply, Honorius of Autun: "There is no other authority than truth proved by reason."

In the thirteenth and early fourteenth centuries, science reached its medieval maturity. It began to amalgamate its three or four tiers: religious wisdom, secular learning, technology, and a pinch of magic reacted upon one another, both because new findings in each tier caught the attention of specialists in the others and because the best researchers became experts in all four. The object of science, too, became broader and deeper: new problems, new fields of research, new countries were explored; personal experiments and firsthand observations were carefully reported; the encyclopaedic collections of unrelated and unchecked lore that had been the delight of countless generations gave way to methodic summations of updated learning. The audience of science swelled: it was taught to large university classes and vulgarized in books for a growing number of literate laymen; it trickled down to common people in simplified and often distorted form. Science developed as an independent profession, exercised by laymen as well as by ecclesiastics: protected by the powerful, patronized by the rich, respected by entire communities, a body of astrologers, alchemists, engineers, and other full-time researchers and consultants were no more insensitive to honors and money than their modern counterparts, but on the whole practiced their craft with love undeterred by fear.

There was still no safety in scientific pursuits, nor would there be for centuries; one could easily end up in jail and burn at the stake on charges of magic practices or heresy. Though outright magic was positively damnable, just a pinch of it hardly attracted attention; the Middle Ages did not yet suffer from the chronic witch hysteria that came with the Northern Renaissance, and seldom mistook a scientist for a sorcerer. Heresy, however, was in every sense the most burning religious problem of the thirteenth century, and scientists were likely targets of inquisitorial proceedings because of their views and wealth. Astrologers who, like psychoanalysts today, were made conspicuous by their wide and important clientele, were especially vulnerable on one point: granted that God in his wisdom had endowed all celestial bodies with an influence on the temperament and destiny of people, it was indispensable in every horoscope to make room for the free will of the individual and his capability to bend in his favor the original thrust of the stars. Not a professional astrologer, Dante carefully stressed this point in the Comedy; he also protected himself by claiming the indulgence due to allegories of poets, who, unlike straight philosophers, may use fables to popularize obscure truths. But his rival, Cecco of Ascoli, poet and astrologer, rejected all compromises, practiced divination without reference to free will, and got away with his pride—until he committed the faux pas of forecasting that the little daughter of his patron (the future Queen Joan I of Naples) was star-bound to become a lecherous woman. The horoscope turned out to be accurate, but Cecco, abandoned by his patron, was burned at the stake.

Nevertheless, it is obvious that by the thirteenth century, the pressure of public opinion had forced the Church to shed most of its a priori Augustinian diffidence towards human reason and science as a "knowledge of human, temporal things." Indeed it would have been preposterous to reject a *scientia* whose basic assumption was that God's will was the explanation of everything, and whose final goal was to read in the natural order the expressions of that will. Moreover, the growing collaboration between lowly mechanics and loftier science made the latter more readily acceptable; for manual labor tempered whatever pride might be involved in abstract scientific speculation, and its usefulness as well as its humility had always placed it above suspicion. There was a message in the well known legend of the acrobat who, having nothing else to offer Our Lady, spent a whole night doing his despised tumbling tricks in her honor and was rewarded with Paradise. As is pointed out by Lynn White, the best advocate of medieval technology, in the early Middle Ages already the Western Church diverged from the Greeks, the Muslims, and the Jews by accepting in its liturgy the organ, a mere machine, as a prop for the praying voices. By the early fourteenth century, the newest models of astronomic clocks were welcomed inside churches, "less to tell time than to demonstrate visually the orderliness of God's cosmos." One may add incidentally that, at the same time, Italian merchants were bringing to the Far East textile goods and at least one clock in exchange for Chinese raw silk and Indian pearls: the first indication that the combined progress of technology and science had made Europe an exporter rather than an importer of industrial products, as far as the end of the known world.

What about the achievements and failures of medieval science? They cannot be described in a short time, and they are not relevant to the theme of the conference. What is relevant is that, by the thirteenth century, a class of professional scientists had emerged, and that public opinion trusted their efforts while fearing their occasional excesses. It is not fair to call them quacks or pseudo-scientists because a large proportion of their theories were based on wrong assumptions; we do not yet know how many wrong assumptions vitiate the theories of modern science. At least we must be grateful to them for their courage, and for the immensely useful inventions they picked up along the way. Mechanical clocks, eyeglasses, alcohol (not an import from Islamic countries, as is popularly believed, but an independent invention of Western alchemists): let the one who despises medieval science learn to get along without them.

BIBLIOGRAPHIC NOTE

The study of medieval science and technology is one of the youngest branches of total history; its bibliography is both overwhelming and full of gaps. The following list includes only a few recent works which have been useful to me and contain supplementary bibliographic indications; Crombie's and White's can best serve as introductions to the problems. I may add that I have developed at greater length some of the suggestions of the present paper in two of my earlier books—*The Birth of Europe* (New York: Evans–Lippincott, 1967) and *The Commercial Revolution of the Middle Ages* (New York/Cambridge: Cambridge University Press, 1976)—and that I might never have fully realized the importance of the field if I had not read the pioneering symposium on "Les inventions médiévales" in the *Annales d'Histoire Economique et Sociale*, VII (1935), directed by Marc Bloch.

M. D. Chenu, *Nature, Man and Society in the Twelfth Century* (Chicago: University of Chicago Press, 1968).

C. M. Cipolla, *Clocks and Culture* (New York: Norton, 1978).

A. C. Crombie, *Medieval and Modern Science* (2 vols., Garden City, N.Y.: Doubleday-Anchor, 1959).

R. C. Dales, *The Scientific Achievement of the Middle Ages* (Philadelphia: University of Pennsylvania Press, 1973).

M. T. d'Alverny, "Astrologues et théologiens au XIᵉ siècle," *Mélanges offerts à M. D. Chenu* (Paris: J. Vrim, 1967).

M. Daumas, ed., *Histoire general des techniques* (Paris: Presses universitaires de France, 1962).

R. J. Forbes, *A Short History of the Art of Distillation from the Beginnings up to the Death of Cellier Blumenthal* (Leiden: Brill, 1948).

E. Grant, *Physical Science in the Middle Ages* (New York/Cambridge: Cambridge University Press, 1971).

V. Ilardi, "Eyeglasses and Concave Lenses in Florence and Milan," *Renaissance Quarterly*, XXIX (1976).

J. Le Goff, *Les intellectuels au Moyen Age* (Paris: Éditions du Seuil, 1957).

E. Peters, *The Magician, the Witch, and the Law* (Philadelphia: University of Pennsylvania Press, 1978).

E. F. Rice, *The Renaissance Idea of Wisdom* (Cambridge, Mass.: Harvard University Press, 1958).

M. Simonelli, "Dante's Usage of Allegory," *Dante Studies*, 1973.

P. Sternagel, *Die Artes Mechanicae im Mittelalter* (Kallmuenz/Opf.: Lassleben, 1966).

L. White, *Medieval Religion and Technology* (Berkeley/Los Angeles: University of California Press, 1978).

P. Wolff, *The Awakening of Europe* (Harmondsworth, Middlesex: Penguin, 1968).

Chapter 4

Discussion of "Wisdom, Science, and Mechanics"

Summarized by Andrei S. Markovits and Karl W. Deutsch

Robert S. Lopez followed the presentation of his paper with concluding remarks. He first described how "the gap of technology" was the greatest gap in the support of science by medieval society. "Technology lived as best it could," he said. "In fact, had it not been for the rise of cities, it might have never gotten off the ground. The practical laboratories of the cities provided the only real technological progress." From the cities came eyeglasses and the study by philosophers of magnetism; the development of magnets was tied intimately to the progress of navigation.

Technology was communicated underground, so to speak: it was not official and it was not protected. For example, the best medieval maps were popular sailors' maps from Portugal. (Indeed, Lopez stressed, these maps were not surpassed until the nineteenth century.) Though such discourse was not officially sanctioned, neither was it forbidden in the open forum of the cities. Dante was allowed to speak on physical subjects to philosophers in the universities, Lopez said, "not because he was a 'laureat philosopher,' but because he was a citizen, and in the cities whoever had knowledge was heard."

The shortcoming of science then, in the eyes of its contemporaries, was the sin of pride. But Lopez noted that such a sin by science is perhaps still punishable today. His analogy: Einstein, too, was a philosopher of the cosmos; he wanted to find a single principle, the quintessence of everything. He sinned of pride but once, when, fearing that the Nazis would discover the atomic bomb before the Allies, he encouraged nuclear research. "I still think he was justified," Lopez said, completing his remarks, "but we must remember that the real problem for science if science is to be loved and not feared is to remember that *alle Menschen werden Brüder.*"

I. Bernard Cohen opened the discussion by asking to "add a footnote" that there seemed to be a gap also between most of the scientists described by Lopez and "anything experimental anywhere." The *calculatores* in several cities, for instance, worked out the laws of uniform accelerated motion but never applied them to real motions. In fact, only by accident did anyone do this before Galileo. Other mechanical innovations—for example, the wheel and the harness—were neither produced, influenced, nor even discussed by scientists. Cohen cited one commentator, who noted that the reason that anatomy and medicine had fallen into such low regard was that the doctor was an academic, "a philosophical type of man," who left anatomy to the barber surgeon, a craftsman. The craftsman was on a lower level of society, out of touch with scientists, no matter how successful he was at his craft. Galileo tells us how he had come upon some men with a pump, Cohen related, "and found that the pump could not suck water up more than about thirty feet, which every pumpman knew. To Galileo, this was a surprise."

More needs to be known, Cohen said, about cases in which the science of that time touched upon the practical, such as in alchemy. Alchemy was satirized by Chaucer in *The Canterbury Tales* as a field that "explains the unknown by what is even more unknown," and Cohen wondered whether this ridicule accurately portrayed the general reaction in medieval times to alchemists and perhaps other scientists, and whether, if we want to comprehend the entire picture, we should go beyond the study of scientists, who were "removed from any practical influence and were of importance to society only insofar as their ideas touched upon general philosophy, theology, and statecraft."

Lopez answered these concerns with a yes and a no. There was, as Cohen suggested, an emphasis on mental over manual activity, though both were needed in science. The mind was often misleading, but that went back fundamentally to Aristotle, who had taught to ask not "how," but "what for"; the scientist is concerned with causes. For example, in Einstein's "cosmic religion," the world has causes, but it does not have a purpose. As for the influence of medieval scientists, Lopez held that scientists did indeed have great influence. Astrologers, for example, went from court to court giving advice to royalty.

Karl Deutsch then posed two questions. First, to what extent was the stress of the early Middle Ages concentrated on the importance of applying knowledge to good ends, in response to the experience of collapse following the fall of the Roman Empire? Today we wonder about the uses of science to rebuild the world if our civilization was to collapse with nuclear war. What was the relation between science and recovery?

Second, to what extent did science gain from the increase in learning from other civilizations? The Muslim, the Chinese, and the Central Asiatic civilizations had not collapsed as had Rome. When Marco Polo went to Peking, he saw a city that was thriving culturally and that dwarfed medieval Rome. The Arabs had boats that could sail against the wind; the Mongols had stirrups superior to those of Europe. Finally, gunpowder and clockworks: were these too not borrowed? Was not Europe's greatest advantage its amazing capacity to learn from other cultures, and was not its adaptation of their inventions to its own purposes, rather than its own scientific achievements, which helped hasten its arrival into the modern era?

To Deutsch's first question, Lopez replied that the collapse of civilization did indeed create mistrust in science, but that, even then, it was a mistrust of technology. The Roman aqueducts, designed for water service as useful testimonies to man's ingenuity, came to be seen instead as useless pyramids, like the great but unproductive monuments of the Greeks.

On the second point, Lopez postulated a yes, with qualifications. He noted first that the Europeans borrowed more from the Chinese than they did from the Arabs, though Chinese civilization was geographically more distant. "But the Chinese did not," he insisted, pointing humbly to his own research and documentation on a matter of particular interest, "contribute spaghetti." He had conclusive evidence that it existed in Italy before Marco Polo went to China, which is the point at which the introduction of spaghetti to Italy is commonly placed.

Therefore, all joking aside, the linkages are difficult to measure. The Chinese certainly had much greater technology, but the general fact that the Chinese knew no Latin and the Europeans knew no Chinese hindered its transmission. The Chinese made great advances in clockmaking, for instance, but it can be held that the Italians themselves were the ones to make the important, final strides in European clockmaking. Progress might have been hastened, however, if discourse between cultures had been wider. The contributions of the Arabs, in contrast, were much smaller. On the question of gunpowder, there are missing links. The Byzantines had (in modern terminology) flamethrowers, but they seem to have forgotten their use until the Chinese reintroduced gunpowder.

However, whether it was really the Chinese influence is unclear; gunpowder seems to surface from nowhere, as it were, in the thirteenth century.

To the general concern whether astrology or folk craft, for instance, constituted science, and whether it is enough to recognize an organized body of men in a professional context as scientists, Lopez pointed out that there is a difference between what we today consider science and what was viewed as science in medieval times. "I do not believe that astrology cannot be regarded as a science," he maintained, "because what matters is curiosity . . . and this is what the religious strictures [of the time] tried to repress. So long as there is curiosity, there is learning." True, such scientific discovery may be accidental, but eyeglasses were invented with principles discovered by someone who was looking for something else. This invention, in the form of navigational eyeglasses, helped Europe to conquer the world.

Lopez concluded the discussion by asserting that technology must be guided [by science], but that science, too, must be guided, and that it is only by the real fusion of the two that mankind can advance. "Fundamentally, and here I am again in agreement with Aquinas," he stated, "I think the important principle is love; if we love, we cannot go wrong."

Chapter 5

The Fear and Distrust of Science in Historical Perspective: Some First Thoughts[1]

I. Bernard Cohen

When I began to reflect on the general theme of this conference, the fear of science and the trust in science, the first image that came to mind was the pact between Doctor Faustus and the devil. In a very real sense this characterizes a point of view that has long been with us. The Faustian story arose and became widely known in the first centuries of modern science and was kin to the general belief that science is itself a black art, a kind of knowledge that is obtained illicitly, and one that has obvious links with magic. The scientist as magician is an image that arises because the experimenter is engaged in unwonted tampering with nature; he appears to be a menace because he aims to unlock forces of nature which should be left as they are. It is conceived that the scientist's payment for the gratification of his curiosity about the nature of the world and man is a Faustian bargain, entailing the loss of his immortal soul: a pact with the forces of darkness rather than those of light. We may note in this context that when Francis Bacon prophesied that science would be a benevolent force, declaring that science would make the world more comfortable and its goods obtainable with less labor, he did not say that science would be a moral force in the community. But no doubt he was aware of the charge that science was allied with the forces of darkness when he invented the concept of "messengers of light," those whose task it would be to go out in the world from the central scientific organization in order to bring the knowledge and benefits of science to men and women in underdeveloped countries and regions.

The fact that alchemy was part and parcel of the "new science," and that another component of it was "natural magick," tended to reinforce the notion of experimental science being allied with the forces of darkness, or magic. The older image of the elderly scientist as magician, with his long coat covered with stars and his conical cap, still lingers on with us, although nowadays transformed into the picture of a young man in the white jacket of the medical scientist or the laboratory coat of the experimental scientist; he continually assures us in television advertisements and the printed pages of newspapers and magazines that the product for which he stands is good for us. Only the little conical cap and perhaps the stuffed crocodile hanging from the roof are missing. In the early days of modern science, alchemy was a force of darkness in a very real sense. It has been suggested that the word "alchemy"came from the Egyptian word for "dark" or "black" or "black earth" and either referred to the black earth of Egypt, the legendary home of alchemy, or the nature of the subject, the black art. The legacy of this beginning was long-standing. In the sixteenth, seventeenth and eighteenth centuries, and until well into the nineteenth, the chemical laboratory, which eventually superseded the alchemical laboratory, was built underground, deep in the basement, under groined arches of stone. In some colleges and universities, the only entrance to such a laboratory might be through a trap door in the floor.

The tinge of magic and the black art is not the only sign of a kind of fear or distrust of science going back to the time of its birth. We may assume that modern science began in the sixteenth century, and we may take 1543 as a symbolic date. This is the year in which Vesalius published his book on the fabric of the human body and Copernicus his treatise on the revolutions of the celestial spheres. Both of these men aroused feelings of great hostility. In the case of Vesalius, the hostility arose from the fear and abhorrence of the human body being desecrated by anatomists, a state of mind abetted or reinforced by traditional religious prohibitions of dissection of the human body. This fear—I may add—has long been with us; in the late nineteenth and early twentieth centuries it gave rise to a whole technology to thwart graverobbers seeking corpses for medical students. Not only was the anatomist someone to be feared and distrusted, but in his role as physician the anatomical scientist aroused distrust. To me there is something more than merely symbolic in the fact that Vesalius was alleged to have been punished for having shown such apparent contempt for life that he performed an anatomy on the body of a patient who was still alive:

displaying supposedly the attitude of the anatomist-scientist toward a patient considered as a potential corpse for dissection and study. To do penance for this crime he was forced, at the height of his career, to go on a pilgrimage to the Holy Land, from which he never returned, perishing in a shipwreck. It is also well known that Copernicus's ideas aroused great hostility; he was castigated by Luther and Melanchthon, even before his book was published, on the grounds that he was a dangerous fool seeking to overturn the universe. Luther explained that the only way to know about the universe was through holy writ and not through science. Having heard that Copernicus would place the sun at the center of the earth and put the earth in a motion, Luther replied that according to Scripture it is the earth that is placed at the center and the sun that moved around it. Melanchthon called for some Christian prince to restrain this man Copernicus. As the historian J. L. E. Dreyer remarked after studying this issue, Melanchthon's words were finally heeded by the Roman Catholic Congregation of the Inquisition and hence Protestants have no right to criticize Catholics for their rejection of Copernicus.

It is to be noted that the official condemnation of Copernicus by the Church alleged that the Copernican doctrine was heretical and philosophically absurd. It was dangerous nonsense; but it was condemned not so much because it was nonsense but, more to the point, because it was dangerous and heretical, and because it jeopardized sound faith and morals. Historians often exaggerate the episode of Copernicus and Galileo and the Roman Catholic Church, especially because there is a mistaken idea that there was some kind of a Copernican revolution attendant on that year 1543. In fact there was no sixteenth-century revolution in the Copernican world nor in that of Vesalius.

It is true, of course, that Vesalius exposed the errors and inadequacies and the faults of the classic establishment of learning based upon Galen, but in doing so he created no revolution. Vesalius himself was a revolutionary figure. In the preface to his great work of 1543, he explicitly condemned medicine for its decline and fall as a result of "the neglect of that primary instrument, the hand," too much having been "relegated to ordinary persons untrained in the disciplines subserving the art of medicine." Vesalius was contemptuous of doctors who, "despising the use of the hands," began to relegate to slaves and servants those "unpleasant duties" of manual treatment of patients and who continued to leave anatomy and surgery "wholly to the barbers." He boldly displayed the new knowledge and showed that much of Galen's teachings must be "false," that Galen had been "incorrect in well over 200 instances relating to the human structure and its use and function." To show that his portrayal of the human body in terms of its fabric or construction was new, based on actual dissections and not on texts of

Galen and other authors, he proudly illustrated the tools of the anatomist: scalpels, chisels, hammers, and saws. Vesalius thus provided the grounds for a revolution or showed the need for a revolution. The reason that he himself did not create a revolution is that he had no system of physiology based on his new anatomy to replace the traditional Galenic physiology. Vesalius, for example, could find no pores in the septum of the heart, through which—according to the fundamentals of Galen's system—the blood must trickle from one side of the heart into the other. He did not, however, invent the revolutionary concept of the circulation of the blood, but could only conclude, "We are thus forced to wonder at the art of the Creator by which the blood passes from the right to left ventricle through pores which elude the sight." In the second edition of the *De Fabrica,* he was a little bolder and expressed his doubt that there are such "channels [as] are described by teachers of anatomy who have absolutely decided that blood is taken from the right to the left ventricle." He still did not fully trust himself "in referring now and again to a new use and purpose for the parts [or organs of the human body]" and so in relation to the heart—he said—"I have brought my words for the most part into agreement with the [physiological] teachings of Galen." Despite any such protestation that "I do not intend to criticize the false teachings of Galen, easily prince of professors of dissection," it was evident to the scientific-medical establishment that Vesalius had made a frontal attack which undermined the foundations of traditional practice. The charge that he had dissected the body of a patient who was not really (fully) dead was possibly related to an accusation of human vivisection of the sort that had been made against other anatomists of that era, such as Berengario da Carpi and Gabriello Fallopio. Such accusations reflect not only the conservative reaction of the traditional medical fraternity against any anti-Galenism but additionally indicate a common fear of anatomy and a distrust of science which obtains its knowledge in illicit ways.

As to Copernicus, the system he set forth in his treatise in 1543 was not at all the one we tend to think of today as Copernican. It did not, in fact, place the sun at the center of the universe; this was rather the achievement of Kepler. Nor did it provide a simpler system, in that it actually used fewer circles than were needed in the Ptolemaic system. Nor did Copernicus completely dispense with epicycles. That also was the achievement of Kepler. From the point of view of practical astronomy, Copernicus attempted to get rid of the Ptolemean "equant" or equalizer, which in many ways was a retrograde step with regard to the problems of positional astronomy or practical astronomy. Thus Copernican astronomy, as a science, did not make anyone uneasy. What was of concern was the general cosmological concept, originating not with

Copernicus but with certain Greek astronomers, that the sun stands still and the earth moves. This concept was discussed or mentioned in the sixteenth century, but not in any really significant way prior to Giordano Bruno. Bruno showed men and women that there were frightening implications of the Copernican general scheme. For if the sun is like a star and is the center of our celestial or planetary world, then Bruno saw no reason why there should not be other worlds, other systems of planets moving about other stars, and thus innumerable forms of life. This was an offensive idea to anyone who held that the earth is unique, not just one out of many, the single body on which God had revealed himself to Moses and the prophets and had made a covenant with the people of Israel. It was to earth and earth alone that God had sent his only son to preach and to be crucified. If one were to believe in other suns and other earths, would it not follow that the drama of the Old and New Testaments would have been enacted an infinite number of times on an infinite number of earths? So offensive was this idea that Bruno was burned to death in the Campo dei Fiore in Rome in 1600 for holding it and other heresies.

When Galileo used a new instrument of science, the telescope, to find that the earth is not a physically unique body in the solar system but is actually more like the planets than different from them, the Copernican problem took on a new dimension of philosophical reality. Here was no longer a proposed system of the world, to be considered as a hypothetical computing scheme, as Osiander had said in the anonymous preface that he wrote to Copernicus's book on the revolutions. Rather, the telescope revealed a new world, one in which the moon is mountainous and looks like the earth, the earth shines and illuminates the moon, Venus shines and shows phases like the moon (and hence must encircle the sun and not the earth), and Jupiter has a set of four moons that encircle it as lesser planets or satellites, just as the lesser bodies of the earth and our planets encircle our sun. If the Copernican world would prove to be the "real" world, in which the earth is merely one of a group of planets which move around the sun, then the literal interpretation of Scripture could no longer be maintained. Galileo's presentation of a physically real Copernican universe was accordingly said to be heretical. It may be noted in passing that while Galileo's book on the two systems of the world was put at once on the *Index Librorum Prohibitorum* and remained there until the 1830s, when the discovery of the annual parallax of the fixed stars by Bessel proved that the earth indeed does move in an annual orbit, Copernicus's book was only put on the *Index "donec corregitur,"* that is, it was put on the *Index* temporarily, "until corrected."

Now the actual corrections were about a dozen in number, and not very fundamental. For example, the prospective reader was directed to strike out the work *proof* in the statement of a "proof that the earth rotates on its axis and moves around the sun" so as to make it read "hypothesis that the earth rotates on its axis and moves around the sun, a proposed demonstration."

Galileo's book was bold, revolutionary, and dangerous (and it was written in the vernacular, without mathematics, for all to read), and so it was proscribed; but Copernicus's treatise (written in Latin, "for mathematicians"—as Copernicus said in the preface) was not dangerous to faith and morals. I believe there can be little doubt that for many fundamentalists, the new astronomy inaugurated by Galileo's telescope was threatening and dangerous, not just because its conclusions contradict Scripture, but rather because it suggests an alternative truth to that of Scripture, a secular truth learned from studying nature and not by studying Scripture or the writings of the doctors of the Church or the inner light. In other words, I believe that the new science of astronomy caused fear and aroused antagonism because it proposed a new and better way of "knowing" about the world, and so shook the foundations of traditional religion, morals, and codes of behavior.

In short, Galileo was advancing a new secular epistemology. He was not only in revolt against established authority—he was also preaching revolution.

* * *

Thus far, I have been stressing certain features of the first century of science which caused fear and distrust, whereas it is more usual to emphasize the promise that science would produce useful things to change men's lives. We are all aware today that in the seventeenth century some Protestants sought to teach that one should have trust in science and not fear science, on the grounds that science is a way of learning about God by studying his creation, the world of nature. Thus science was conceived to be truly "useful," because it serves religion. A typical representative of this new point of view was Robert Boyle, a mighty corpuscularian and a follower of the philosopher Francis Bacon. Like many of his contemporaries, Boyle sought to reconcile science and the Protestant religion by essentially proclaiming that there can only be one truth, made accessible to man equally through divine revelation and through the study (by experimental science) of God's handiwork in nature. In other words, sermons are to be found in "sticks and stones." In thinking about this seventeenth-century position, however, we cannot ignore the fact that scientists and their spokesmen were also engaged in loud propaganda that the new science would be of direct practical benefit to man. Trust us, the scientists were saying, and we will improve the methods of mining and manufacture, the modes of transportation and communication, the practice of agriculture, the way in which we attack our enemies and defend our homes and our land, and we will provide a new basis for medicine which will conquer disease and give us all longer, healthier lives.

This doctrine, proclaimed early in the seventeenth century by both Francis Bacon and René Descartes, who were its chief spokesmen, may be called the doctrine of practical improvement. It was argued strongly enough to cause Louis XIV, through his minister Colbert, to spend a considerable sum of money to establish an observatory and an Academy of Sciences in Paris and to hire academicians to advance the experimental sciences for the greater glory of the economy of the realm. In England, no such royal funds were made available for the Fellows of the Royal Society of London nor for the institution itself, but there was created— a little over three hundred years ago—the first large-scale institution of science to be supported by government, the Royal Greenwich Observatory, among whose aims was to improve navigation and, in particular, to discover a simple means for determining longitude at sea.

To me it has always been a source of great fascination that scientists were so successful in marketing their promised practicality that governments were willing to write fairly large checks, and to continue to do so— even though, as the years went by, there was no immediate prospect of a substantial payoff. It is a fact of record, and it is important to make that clear in the present context, that from the middle of the seventeenth until well into the nineteenth century, scientists kept saying that advances in fundamental scientific knowledge or "pure" science would lead to practical benefits and innovations of use to man, but that there was no dramatic major alteration in the way in which men earned their living, grew their food, transported themselves and their goods, communicated with one another, attacked their enemies and defended their countries, cured diseases and protected their health.

Until well into the nineteenth century, there was only one notable instance of research in pure science which led to a practical embodiment of this kind. That was Benjamin Franklin's invention of the lightning rod around 1750, an invention that was based on his discoveries in the pure science of electricity, including the nature of grounding and insulation, the action of pointed conductors when grounded and insulated, and the nature of silent and spark discharges, plus the applications of all these findings to prove in the first instance that lightning *is* in fact an electric discharge. As late as the 1830s, some two centuries or more after the original doctrine of utility had been proclaimed by Bacon and Descartes, the French astronomer Arago—like other scientists of his day —still had to ground his request for monetary appropriations for the support of science on the premise that disinterested research would some day lead to something useful or practical. When pressed, during the debate in the Assembly, to give an example of any significance to document the claim that the advance of knowledge produces something useful or of benefit, Arago could point only to Benjamin Franklin's invention.[2]

I cannot mention the invention of the lightning rod without indicating some of its consequences for the main theme of fear of science and trust in science. One of the things that Franklin's discovery of the electrical nature of the lightning discharge accomplished was to strike another blow for rationality and the decline of superstition. Until that time, and even for some time afterward, there was a general belief that bolts of lightning were sent by an angry God, who either wanted to punish sinful man or at least to give him a sign and warning of His displeasure at the way in which man was behaving. In fact, in Franklin's day, it was quite customary to ring church bells during severe lightning storms, something which tended to make bell-ringing quite a bit more hazardous an occupation than it is now. Church bells then tended to bear an inscription cast into them, reading:

Vivos voco, mortuos plango,
Deum laudo, fulgura frango.

That is, "I announce births and I mourn deaths, I praise God and I free the air of lightning." Some in the audience may remember these lines, which appear prominently in Schiller's poem "Die Glocke." Franklin's discovery showed that far from being a sign of divine wrath, a lightning discharge occurs when there is built up an opposing charged cloud and charged region of the earth, according to a process which we now call electrostatic induction. Not only did Franklin show that the lightning discharge is merely a natural phenomenon, but at the same time his invention of the lightning rod showed man how to circumvent the destructive effects of lightning and to safeguard our homes, our barns, and our churches.

You might suppose that since Franklin's invention occurred in the middle of the eighteenth century, which we tend to think of as the age of reason and enlightenment, the lightning rod would have been universally hailed and at once put into general use. Far from it! As a matter of fact, even to this day lightning (at least in America) remains high (sixth place) in the list of the causes of death and destruction by fire. It is certainly clear that the lightning rod is not being used to the full extent that its practical aspects justify.

In the eighteenth century, many ordinary people objected to the introduction of lightning rods, and so did some men of learning. Generally speaking, the reactions against the use of lightning rods show us the essential tension between science and the public at large, the component of distrust in science or in scientists. A typical anti-lightning rod reaction occurred in Boston, where it was argued that if a large number of iron points are erected in the air, they may indeed circumvent the production of destructive strokes of lightning, but of course they cannot prevent the ultimate destruction intended by God. Accordingly, the lightning will accumulate in the earth and produce an earthquake (another sign of God's wrath). In Austria, where a Roman Catholic priest made a kind of partially independent invention of the lightning rod, the local farmers tore it down since they believed it had caused a drought. There were many other cases in the eighteenth century, the most famous of them all associated with a Monsieur de St. Omer, who insisted in keeping up a lightning rod despite the angry protestations of all of his fearful neighbors, who were sure that the rod was inviting death and destruction. The reason that this case is so famous is that it launched the career of a successful young lawyer, Robespierre.

It should be added that there are many instances in the nineteenth century, as well as in the late eighteenth, in which there has been exhibited a great fear of lightning rods—perhaps equal to (and maybe even greater than) the fear of lightning itself. Psychoanalysts have pointed out that such a reaction to lightning is associated with fear of the father, possibly in association with the image of God as an angry father hurling thunderbolts at his naughty children. Using a lightning rod would then be a dangerous thing to do since it is an attempt to circumvent a father's wrath.

The discovery of the electrical nature of the lightning discharge was not the first blow struck by science against a superstitious fear of natural phenomena. One of the fruits of Halley's analyses and Newton's celestial mechanics was the concept that comets are—as Newton put it—"a kind of planet" and so they (or at least many of them) move in elliptical orbits. The importance of this finding is that many comets must return at regular intervals to the central regions of our solar system. One of the most famous of all comets, "Halley's Comet," has a period of about seventy-five years. Those of us who live at the present time apparently can have no sense of what the feeling must have been like in the sixteenth, seventeenth, and eighteenth centuries when Halley's Comet appeared and reappeared. It hung in the sky at about the same place for days and days, covering a large area of the heavens, and emitting light of a dull and somewhat dirty yellow color. It was only natural that people would then ask themselves: What does it mean? What does it foretell? Why has God sent it, or what is the message that God intends by it for man? Indeed, such unusual aspects of the world of nature were known as "divine providences," or messages of just this sort. But if the comet has come back every seventy-five years and will continue to come back every seventy-five years, then its appearance is not an extraordinary but an ordinary natural event; the comet does not carry a special message for you and for me.

*   *   *

The incidents of the lightning rod indicate what I see as a general theme, namely, the fear of the dangerous outcome when a scientist tampers with nature. This is manifested in the public reaction to the new eighteenth-century practice of inoculation for the smallpox, a procedure in which the "matter" of a mild case of smallpox is inserted directly into the human body. The question at once arose: Should one interfere with the scourge of smallpox visited upon man by his God? I have phrased this in a religious vein, but the sentiment at issue is the distrust of the scientist when he tampers with the order either imposed on nature by God or by God's actions through nature, and therefore is very relevant to our theme. Many people in the eighteenth century believed that inoculation would spread the disease rather than prevent an epidemic, thus expressing the kind of distrust to which I have been referring.[3] Do not tamper with Nature!

This theme appeared in a dramatic way when I was a young boy. The Schick test (for diphtheria) was just then being introduced in the schools, and children had to get the permission of their parents before the test could be given. I well remember that all of us ten- and eleven-year-olds were convinced that the test would produce a partial or complete paralysis. For days after I was given the test, I kept waiting for the dreadful signs to appear, sure that this was the price I was to pay for having suffered this dreadful introduction of a foreign substance into my body.

*   *   *

It is one of the unfortunate aspects of intellectual history that we generally have to be confined to the evidence of written documents, which means that we tend to deal with the thoughts of literate men and women and not those of the nonliterate members of the population. For the most part, the ideas and opinions we study are those of literate men and women who were willing to express abstract ideas and thoughts on paper. We have no way of telling very much about what the ordinary men and women of the eighteenth century thought about science. Probably most of them did not know much about it, and cared less. They would be apt to encounter science only through contact with a particular action of a scientist. By this I mean that they did not tend to encounter the abstract theories of science or the newest scientific discoveries or the observation of remarkable natural phenomena, although they might encounter a lightning rod, or find that a new means of preventing or curing a disease had been introduced.

Of course, an exception must be made for that group of Protestants who would hear their clergymen preach sermons (for the reason mentioned earlier) that included many of the new discoveries and phenomena of science and even scientific theories. From such sermons, as Perry Miller remarked, it is easy to compile an encyclopedia of the scientific knowledge of that day. However, in these cases, although we may know the science to which ordinary churchgoers may have been exposed, we have no way of gauging the effects such exposure to science would have had on the members of the congregation. We cannot take polls or accurate counts, as allegedly we can do today, in order to determine what the state of public opinion was in the eighteenth century —or even in much of the nineteenth century—with regard to science, scientists, and particular innovations. All that we can do is look at certain samples, as I have been doing here, and try to use these as indicators of a general feeling of a certain kind.

* * *

It is not entirely clear from the general statement of the purpose of this conference as to whether the theme "fear of science" was also intended to include antipathy toward science, or at least antipathy toward scientists. I find the latter theme to be not unrelated to the general one that we have been exploring. In the sixteenth and seventeenth centuries, the scientist was often mistrusted and disliked, as even in the eighteenth century, because he advocated a path to knowledge alternative to the traditional one based upon God's revelation; he was a man who tampered with the forces of nature and tried to oppose God's will; and he was someone who was making dangerous experiments which might have disastrous consequences for us all. However, I must remind you that in the sixteenth, seventeenth, and eighteenth centuries, science was to a large degree impotent. Science was feared for its potential rather than its actual power over nature. By this I mean that science was not as yet the fount of great changes in our way of life or our conditions of health and living that we have come to expect today. Accordingly, the examples of active distrust and antagonism tend to be minimal in comparison to the large body of satire directed against this small intellectual company of devoted men and women who were engaged in such apparently ridiculous pursuits as weighing the air.

There are many plays and essays poking fun at science and scientists. *Cyrano de Bergerac,* you will remember, presents a series of absurd devices by means of which a scientist might go to the moon. Swift satirized the scientific academies in his *Voyage to Laputa.* Shadwell's *The Virtuoso,* a popular play of the late seventeenth century, made fun of a scientist who knew everything, a character in many ways like Robert Hooke, a real virtuoso. Butler wrote a devastating account of a group of scientists looking through a telescope and discovering "An Elephant on the Moon," which proved to be only a common louse which had been walking across the objective lens of the telescope.

Ever since the seventeenth century humanists as a group have generally been distrustful of scientists, and envious of them too—resenting the fact that scientists have always been able to get financial support for their research and other perquisites which are denied to classicists, historians, and literary men and women. I do not know how old this feeling is, but it certainly expresses itself in literary satires of the kind that I have mentioned, going back to the seventeenth century. Further, it appears in a simple and dramatic form in the complaint of the humanist members of the Paris Académie des Inscriptions et des Belles Arts, who were very much concerned by the fact that their opposite numbers in the Académie des Sciences received larger annual stipends than they did. This is a recurring theme, rearing its head in a famous letter of the nineteenth century, written by Lewis Carroll to the London papers, complaining that Oxford scientists were getting support in the universities to a degree denied to humanists, and that this was money wasted or badly expended.

* * *

The eighteenth century saw the rise of a great public interest in science, as may be seen in the invention of special instruments to explain science to the layman. One such was the orrery, or mechanical planetarium or model of the solar universe, which was as much a feature of popular scientific entertainment in the eighteenth century as it is today. The eighteenth century witnessed the formation of great collections of natural curiosities, museums of natural history, and gardens in which the knowledge of plants and their distribution was set forth in a scientific fashion, such as according to the scheme of Linnaeus. There were many popular itinerant lecturers on science, who traveled far and wide to entertain the curious and to show the latest novelties of science by means of experiments and demonstrations. Sometimes, such sets of lecture demonstrations were institutionalized, as in London, where they were given by Whiston, Desaguliers, and others; and in Paris, where the Abbé Nollet became a sort of official demonstrator in science for the members of the Court of Louis XV.

Diderot's *Encyclopédie* lauded the sciences and their achievements to a large group of readers.[4] Furthermore, general magazines of the eighteenth century devoted space to explaining aspects of science and describing the latest scientific discoveries, along with the customary disquisitions on history, poems, accounts of battles and historical events, and political news. Such information even found its way into the newspapers and became the subject of conversations in the coffeehouses. Thus we can see many signs of a growing general interest in science and its achievements that becomes particularly marked from the middle to the end of the eighteenth century.

This is the famous age in which the notion of progress becomes associated with the sciences, as well it might have been, for it was clear that in the sciences there was a meaning to progress which just did not exist in any other field of endeavor. Milton, of course, was a great poet, but did it make any sense to compare his greatness to that of Homer? Was Locke truly as great a philosopher as Plato or Aristotle? How different it was in the case of Newton and Archimedes, for no one conversant with the *Principia* could doubt that Newton knew much more physical science than Archimedes himself. Not only did science give a model of the kind of progress that men could hope for in society and in human relations, but it also came to be thought that the advances in the sciences and their eventual applications would improve man himself.

Such improvement was even conceived in measurable terms—an indefinitely long life span, freedom from diseases, and so on. Turgot was but one of many who developed this theme, but it reached its high point in the eighteenth century in the *Esquisse* of Condorcet, who saw in the development of man a series of stages similar to the stages in the history of the earth envisaged by Buffon, each representing a step of progress related to science. In the end, thanks to science, not only would the state become more perfect and society itself more equitable, but even death would be so far removed that it would be looked upon as a curious accident. This almost eternal life and happiness was to be the result of science. Here, then, toward the close of the eighteenth century, is the supreme statement of trust in science and the benefits that must come from the works of science.

* * *

And now we come to the nineteenth century. Just as we may consider the *Esquisse* of Condorcet the final statement for the eighteenth century of infinite trust in science, so—in the beginning of the nineteenth century —may Goethe symbolize for us the old distrust in science which lingered from the beginning days of the scientific era. In more powerful literary and imaginative terms than ever before, Goethe set forth anew the concept of the Faustian bargain made by the scientist with the devil as the price of obtaining deep, inner knowledge of nature.

I well remember, as many of you may also, the strong impression made upon me by reading Goethe's *Faust*. I cannot tell you what a shock it was to me later on to discover that Goethe was not just another antiscience humanist, but that he actually was a man of science as well. Not only did he write in a deep and penetrating manner about many aspects of plant science and natural history, and even the development of man and his species, but he also made physical experiments on light and discovered the intermaxillary bone. Goethe, as is well known, concluded his optical researches by attacking the simple Newtonian theory of light and colors. In this he was not merely being anti-Newtonian or adopting an anti-establishment position. Rather, he was concerned with the psychological and physiological aspects of color vision and perception and not just with the physical light.[5]

At this same time, the beginning of the nineteenth century, there was another current of thought which also opposed the Newtonian explanation of light and color. This came from a group of poets and artists, at least in England, who felt that Newton had destroyed the mystery and beauty of the rainbow with his analysis of sunlight by means of a prism. In drinking a toast "to the confusion of Isaac Newton," these artists and literary men were expressing a basic antipathy to scientists, whose cold analytical minds threatened to destroy the simple and direct appreciation of nature by artists and poets. Here, then, we have the positing of an antipathy, which is widespread today, between the aesthetic contemplation of nature and its beauty on the one hand, and the hard, analytic, corrosive analysis by science on the other. This opposition was also expressed by the visionary William Blake who, in his hatred of science (particularly Newtonian science), wrote the lines:

> The atoms of Democritus
> And Newton's particles of light
> Are grains of sand upon the shore
> Where Israel's tents do shine so bright!

This was an expression of the same kind that Blake gave vent to in the poem, reading in part:

Mock on, mock on, Rousseau, Voltaire.

* * *

In the nineteenth century, it is easy to find fear of science and its expression as opposition or antagonism to science on many levels. One prominent example of the distrust of science in that century may be seen in the practice of medicine. Today we would demand of our doctor that he be up-to-date in every aspect, in touch with the leading current aspects of scientific research, able to give to us the latest and newest medicines. We want a doctor not only to be in direct contact with the world of research, but if possible to be doing research himself. In the early nineteenth century it was quite otherwise. If a doctor—and there are some notable examples on record—happened to be engaged in research, and was interested in new treatments and remedies that were not traditional, his patients would distrust him. They would tend to fear that they might be used as guinea pigs; they wanted to be safe and sure rather than to be experimented upon. In fact, there were medical doctors engaged in research who ended up by having no practice whatsoever. Even as late as the time of Pasteur, the medical profession was far indeed from embracing his new discoveries at once, but rather resisted them— as is very well known. Here, then, we have the fundamental distrust of novelty, which is of course a feature of most of human experience.

* * *

At this point I cannot help but interject a comment about innovation at large. Many people conceive of man as innovative and point to the sequence of inventions which have marked the progress of society. However, I think it can be very well argued that there is a natural distrust of innovation, and that in general people rely on what is "true and tried," the good old-fashioned and well-established ways.

Freud, once commenting on one of the greatest inventions of all, fire, made the following observation. He assumed that this great innovation occurred when a tree was struck by lightning and set afire, and that man learned to tame the fire and to use it only when he was able to overcome his homosexual urge to extinguish the fire by urinating on it. On this supposed great occasion, man tended the fire, maintained it, and learned how to use it. How important this innovation was may be seen not only in the effect it had on man's eating habits, that is, cooking his food. It also gave man the ability to practice metallurgy. In order to win metals from their ores, there had to be set up a special class of society, the smiths, who were not directly concerned with food production. This was possible only with the invention of agriculture, when it no longer required the collective energy of the whole group to gather food. Agriculture led to one of the first great social revolutions in history.

One can see again and again the opposition to innovation, the continuance of doing things in the good old ways. In the Trobriand Islands, for example, as analyzed long ago by Malinowski, there is an original kind of naval architecture, the use of an outrigger, which gives canoes a great stability at sea. It is obvious, when you think about it, that the further the outrigger is from the canoe, the greater the stability; but if the outrigger is put too far out, the sticks that hold it to the canoe will be subject to rupture, and so the effect of safety is negated by the increased danger. There is an optimum size of outrigger for a given size of canoe in terms of the length of the stick that holds the outrigger to the canoe, and this has been worked out and is used again and again. That it is successful is proven by the long ocean voyages that the Trobriand Islanders make. Now the point is that this is a technology that works. Who would want to tamper with it? No one, generation after generation, needs to make experiments. Everything is done in the good, old, tried and true way.

Even today, when we live in a world of tremendous innovation (and one has only to think of transportation and communication and the new materials such as synthetics and plastics to see what an innovative time we live in), there is much that continues in the old way for which the use is long since gone. Take men's clothes as an example. First of all, the jacket has a collar with lapels, one of which has a buttonhole although there is no button to go with it on the other side. The jacket is divided in the back in a way that is most useful if you sit on horseback, although most of us do not use such a "hacking" jacket for hacking. We have buttons on the sleeves (and in some cases even buttonholes) but we never unbutton them. All these are vestigial remains of something that had been functional in the past, when for example the cuffs of a jacket were so tight that they would not slip over the hand and had to be opened—hence the buttons.

Perhaps the most striking example of such vestigial remains is in the motor car. The early bodies for automobiles were made by carriage makers and—as everybody knows—the first automobiles were even called horseless carriages. When the carriage body was transferred from its horse-drawn to its self-propelled function, the little socket which held the whip was still made and put in place. Even when the bodies for motor cars were no longer transformed carriage bodies, but were produced expressly for the new purpose, the little socket for the whip long continued.

As a personal note, I may add that I lectured on this topic in my undergraduate course for many years, and then wondered (as professors sometimes do) if what I had been saying were really so. Accordingly, the next time that I was in Washington, I went to the Smithsonian Institution to look at the collection of antique automobiles. Sure enough, many of them still had the socket for a whip.

Incidentally, until the 1930s the radiator cap was exposed at the front of the hood. To add water to the radiator, it was then not necessary to lift the hood, as one does today, but only to unscrew the cap. This radiator cap was often plain, but eventually various kinds of ornament were placed on it, of which the most famous one is perhaps that of the Rolls-Royce or of the Mercedes-Benz. Today the radiator cap is under the hood, but the ornament is still in place, outside, in full view, and functionless. According to certain experts, the rather inefficient placing of the motor in front of the automobile, with a long drive shaft going back to the rear wheels, rather than being placed at the rear, closer to the driving wheels, is simply another example of lack of innovation. The motor was placed where the horse had always been: out in front.

I introduce the subject of innovation because the fear and distrust of science may very well be associated with the innovative aspect of science. And this leads directly from the theme of innovations in science itself to the dramatic ways in which science, in its practical applications, transforms technology, and also everything else.

* * *

Before getting to that topic, I must mention that the one subject within the development of science that aroused more fear, distrust, and antagonism than any other in the whole history of science—I am now thinking of scientific ideas and theories, not of technology or environmental effects—was the theory of evolution. Of this, I believe there can be no doubt whatsoever. No scientific theory had such an effect upon every thinking man and woman as evolution, with its two aspects: the history of the earth and of life, and the history of man.

The general reaction to these topics was somewhat similar to that aroused by the shift from an earth-centered to a sun-centered universe. This change in the location of the center of the universe required that certain aspects of Scripture be given a nonliteral interpretation, that it be said that the Bible is written in the language of the vulgar and is not meant to be taken at every point as a literal statement or an explanation of facts in the physical universe. Thus when we speak of the sun rising and moving across the heavens, and the like, this is merely an example of using ordinary speech. Even today, when we supposedly have been converted to the Copernican system, we still talk of the sun rising and setting, and so on. In fact, as Galileo argued, using an expression of St. Augustine, it is the purpose of Scripture to tell us how to go to heaven and not how the heavens go.

And so it is with the age of the earth. After all, a century is but a second in the eye of the Lord, and in any case periods of time may not be necessarily the same in Scripture as they are for us today. So the matter of age of the earth is not really a burning issue, except for certain absolute fundamentalists who take their Scripture literally. Moreover, the notion that there has been a kind of gradual evolution of animal or plant species is not of terribly much concern either, except for such fundamentalists. As a matter of fact, many apologists—both Protestant and Catholic—have cited passages from the Psalms and other parts of the Bible to show that a general notion of evolution is compatible with religious belief. In fact, the Bible opens with a discussion of the stages or days of creation itself, so that the idea that everything was not completed at once but required a series of steps is not a hard one to take.

Now I would argue that for most people, or even for almost all people, discussions of the development of the earth and its history, the succession or evolution of animals and plants, and all the rest, are not of much concern—save as they are related directly to man himself. The fact of the matter is that wherever the earthly abode of man happens to be located, whether whirling in space or at the center of the universe, and whether our earth was made at once or took a long time in the forming, and whatever may have happened to the forms of animals and plants that surround us as revealed by the fossil record—none of these findings of science is particularly disturbing to you or me, although (as I have said above) they do antagonize fundamentalists. The scientists' choice was either to believe that one form of animal life had evolved into another, as Lamarck had argued, or that there had been a series of catastrophes, each one requiring new creations, as believed by Georges Cuvier, the great founder of modern palaeontology and the science of comparative anatomy. However, the problem takes on vastly different dimensions whenever such considerations are applied directly to man himself.

* * *

Nineteenth-century literature is full of examples of the reaction to the notion that man is just an animal. In Benjamin Disraeli's famous novel

Tancred, Lord Beaconsfield, the hero, affirms in great anger that "I have never been a fish." In another example, John Fiske records a conversation in which an elderly gentleman said to him in indignation, "I am certainly not a mammal." There can be no doubt that what was chiefly disturbing about evolution, in a way that had never been unsettling in the case of any scientific theory before, was that in its simplest terms the Darwinian evolutionists said to every man and woman: there is a close and direct kinship between you and the animals. The implication of Darwinism for most people was made clear in the title of one of Darwin's books, *The Descent of Man*: man is descended from monkeys or apes.

It will be recalled that Bishop "Soapy Sam" Wilberforce, in his famous debate with Huxley at the British Association meeting, asked him whether it was on his mother's side or his father's side that he was descended from an ape. The bishop was only expressing a fear and antagonism widely held by ordinary people, one that had been aroused by the latest trend of science. In fact, this feeling is still with us.

Those of you who have been reading the writings of Dorothy Nelkin are aware that in California there is a large "creationist" movement which demands "equal time" in the teaching of high school biology for the theory of creation alongside the theory of evolution. It is well known that fundamentalists in America have long tried to prevent the teaching of evolution in schools. There is the famous Scopes trial of the 1920s to remind us that I am not talking about some part of ancient history but rather something fairly recent, of our own century, that is still with us. In the case of the Scopes trial in Tennessee, one could argue that this was a battle between science and some religious fundamentalists or fanatics. However, in the case of the California "creationists," it turns out that those who are leading the movement have scientific credentials. They include engineers and chemists: men who have scientific or technical doctorates. They are not people who can be described as simply ignorant —although there is a sense in which they really tend to be ignorant of certain aspects of the subject that they discuss.[6]

Sigmund Freud was quite aware of the significance of the antipathy of ordinary men and women to Darwinian evolution. Freud saw evolution as one of the three great blows that man has received to his narcissistic self-image. The first, Freud said, was the Copernican revolution, which shifted man's abode from the center of the universe to an insignificant planet wandering around the center. The second was the Darwinian revolution, which showed that man has great kinship with animals, and does in fact have an animal part of him. The third, according to Freud, was psychoanalysis, which showed that man is not completely in control of himself. With customary deep insight, Freud saw that his own revolutionary introduction of psychoanalysis had features in common with Darwinian evolution, at least insofar as the reaction of fear—and hence of mistrust—was concerned.

* * *

Evolution has certainly been of major importance in determining the public reaction to science: a repugnance and fear of a result, and even an antagonism and dismissal of the scientific work, as if it had no sound experimental foundation, no real proof or firm basis of validity, no compelling reason for belief in it. That this has applied equally to evolution and psychoanalysis seems from today's vantage point to be very obvious. On the other hand, it is not quite so obvious with regard to another subject which was developed in the nineteenth century, namely, statistics. This topic is related to the case of evolution and of psychoanalysis in that some people tended to dismiss evolution and psychoanalysis by the phrase "only a theory"—really no more than a way of indicating a fundamental mistrust; similarly, it was (and is) asserted that a statistical proof is no proof at all. It should be added, however, that a number of geneticists and palaeontologists of the late nineteenth century and the first half of the twentieth also tended to reject Darwinian natural selection. Of course they believed in some kind of "evolutionary" change in general, a genetic explanation.

The dismissal of evolution by natural selection or of psychoanalysis by saying "it is only a theory" is paralleled in the realm of statistics. The expression, "It is only a statistical proof," may be taken to mean that since the scientific basis is statistical, the alleged scientific result in question is a doubtful one, one that need not be taken seriously. This is the sense in which "it is only a theory" turns out to be no more than another way of declaring a lack of real trust in the results of science. The statistical basis of knowledge or a statistical explanation seems to many nonscientists either an evasion of ordinary notions of causation or a confession of ignorance. Statistics and probability appear in daily life in relation to doubt in such expressions as "It probably will rain," "Chances are that . . . ," "The odds on Seattle Slew's winning the race are . . ." Each such statement declares a lack of certainty, a degree of error in the prediction. A science based on a statistical model runs up against the conventional wisdom, the dictates of common sense, and general ideas as to the basis of sound and reliable knowledge and the ordering of the world.

For many people, a statistical proof in science appears to be of the same nature as a statistical prediction in "scientific" sampling of public opinion. And the errors in pollstering are, as the lawyers say, "open and notorious." Nowhere is this phenomenon of having no trust in science more obvious than in relation to the question of cigarette smoking. I am certainly not in a position to be able to say whether there would have been a rapid decline in cigarette smoking if there had been found to be a direct and simple causal link between cigarette smoking and lung cancer (and also other diseases, notably of the circulatory system). However, the fact of the matter is that a large proportion of the people who go on smoking cigarettes do so on the grounds that the proof is "only statistical."

I believe that the distrust of statistics, and the consequent rejection of any scientific result which appears to be statistically based, is closely related to the reactions to Darwinian evolution. People disliked Darwinism primarily because it tended to destroy the uniqueness of man as a species among the animals. Whatever evolutionists or Darwinians might say, man is unique, totally apart from the animal kingdom in that he was made in God's image.[7] Similarly, there was a distrust and a fear of statistics because this approach seemed to deny the role of the single individual, who simply became submerged in the mass. Many of the antistatistical people in the nineteenth century argued that statistics is "inhuman." After all, who cares what is happening to the average; everyone is really concerned with what is happening to himself as an individual. Dickens argued against what he called the folly of considering that conditions among the urban slums of a place like Glasgow were getting better because the average condition had risen, when the fact of the matter is that the population had also increased and so there were more people suffering than before.

I do not propose to go here into the pros and cons of statistics, but the argument also ran something like this: who cares about averages and percentages and probabilities when one is concerned that a single person has been killed in a railroad accident? Does a parade of statistics really make anyone feel safer on an airplane? It has been said that one death is a tragedy, a million a statistic. Statistics appears to be even a way of misleading people. Consider, for instance, the phrase, "There are lies, damn lies, and lies with statistics." A fairly recent popular book was entitled *How to Lie with Statistics*. In England, and possibly elsewhere, the reaction against statistics arose in some measure from the fact that the early statisticians, particulary biometricians from Sir Francis Galton to Karl Pearson, were also eugenicists and racists.

*   *   *

The sources of antagonism to science in the nineteenth century go beyond evolution and statistics; there was also the "reductionist" point of view that man is merely a machine, that the functions of the human body can be reduced to chemical and physical actions. While this concept was of great importance for physiology, it was one which aroused an obvious antagonism. In modern times, the notion that animals are machines and that much of man's own action is "mechanical" goes back at least to Descartes, whose "mechanical philosophy" (as it was called) caused such fear that his writings were placed on the *Index Librorum Prohibitorum*, where his *Opera omnia* appear alongside Spinoza's *Opera omnia*. Indeed, so horrendous did the Cartesian notions appear that one of the leading exponents of the Cartesian philosophy, Jacques Rohault, was not allowed to be buried in consecrated ground, although—but for his Cartesianism—he was otherwise a good practicing Catholic. In the eighteenth century, the outstanding proponent of reductionism of this sort was Lamettrie, who wrote a book called *Man a Machine*, and whose works were condemned equally by Protestants and Catholics.

There can be no doubt that in the nineteenth century, the notion that man is a kind of automaton functioning just as a machine was difficult for many people to take. This philosophy "reduced" man to a mere machine in the same sense that Darwin "reduced" man to being an animal. The reductionist philosophy may be compared to the later psychoanalysis, since both argued that many of man's actions are not the result of rational choice and hence are not functionally determined by the mind at the command, independently and with will. Both the reductionist philosophy and psychoanalysis would deny that the character of the human mind and the will derive from the nature of man's immortal soul, a result of his having been created in God's image. I believe that the reaction of fear to both philosophies was very much the same. An additional aspect of reductionism is its direct relation to philosophical materialism, which is in obvious opposition to the established religious value system. Furthermore, radical materialism became closely linked to political radicalism and thus was viewed with suspicion and even fear by the greater part of society.

* * *

I have mentioned earlier the fact that almost from its first days, modern science was dedicated to the notion that it would eventually produce practical innovations of great significance for the life of man. Additionally, I said that up until at least the first third of the nineteenth century, the advancement of scientific knowledge, or what we would call today pure science or basic science or fundamental science or basic research (and it is curious that there is no universally agreed-upon word here) had not produced any notable or major innovation that had made a radical or fundamental difference in the ways in which men grew their food and fed themselves, cured illnesses and prevented diseases, protected themselves against their enemies and destroyed their enemies, communicated with one another or transported themselves and their goods, earned their livings and manufactured commodities, and so on.[8]

The first large-scale remarkable instance of what we consider today to be a primary quality of the scientific enterprise occurred around the middle of the nineteenth century, just about the time of the American Civil War, at the time of the formulation of Maxwell's electromagnetic theory and the time of the publication of Darwin's *Origin of Species*. The area in which this occurred was organic chemistry, notably the chemistry of coal tar and the production of coal tar or analine dyes. Here the world was treated to a remarkable phenomenon, the basic transformation of the life of man by science.[9]

Prior to the 1860s, all materials for dying cloth were obtained from plants and animals. A typical dye was alizarin, or Turkey red, a product of the madder plant. Because of its wide use, this was an important agricultural product along much of the Mediterranean: in Spain, Southern France, Italy, Greece, the Middle East, and North Africa. When alizarin was produced synthetically from coal tar, itself a by-product of the manufacture of illuminating gas from coal, the new dye was not only cheaper and better to use, but more readily available. Within a matter of a decade or more, the whole agricultural basis of this vast region had been radically changed. Furthermore, today a plant which had been commonly grown as the major agricultural production throughout so much of the world has become a botanical curiosity that most of us will never have seen. This was an awesome and impressive example of the power of science.

It must not be thought, however, that men of power and influence universally and at once understood what an extraordinary revolution had broken upon the world. Had that been the case, there would have been an explosive universal development of science and its applications under the patronage and support of government and industry. The fact of the matter, however, is that although the new coal tar industry had begun in Britain, following the discovery of mauve by Perkin as a college student on his Easter holiday, the major action shifted to Germany.

Many of the leading German dye chemists got their initial training in Britain and then went home to use what they had learned there. Year after year, British chemists analyzing the scene deplored the fact that there was no established organic chemistry in Britain and that the progress of the Germans in capturing the lion's share of the world's markets was due to their concern for organic chemistry. However, one must remember that even though Britain was losing the coal tar dye industry to Germany, her own sale of coal tar dyes was continually increasing—for the reason that more dyes were being used all over the world now that they were cheaper. Furthermore, one would not have expected Britain to undertake the same aggressive sort of policy that was pursued by Germany, since Britain was then in the comfortable position of having an empire on which the sun never set, while Germany had none —except what she might be able to win by conquest of science and later by conquest of arms.

A very impressive example of what science could do arose later with the synthesis of indigo in Germany. Here the question was not merely that of producing the synthesis, but of making it practical. Establishing the basis of an industrial production of indigo took a matter of ten years (from about 1890 to 1900) and the expenditure of something like $5 million. It required the concerted efforts of academic chemistry, industrial chemistry, and the intervention of government. This was the first time on record that a "crash program" in science was undertaken with vast government support, in order to achieve a practical goal.[10]

*   *   *

At the same time that science was creating new industries, it was also successful in discovering the microorganisms that are the causes of disease, and in then finding cures. Here it might be pointed out, as an aside, that the coal tar dyes were very important in the development of chemotherapy. Not only was Ehrlich's 606 or Salvarsan a by-product of coal tar dye research, but so too were the sulfa drugs (prontosil) and their derivatives; also, the introduction of Gram staining was a major new tool for the study of microorganisms. The telegraph, based on the nineteenth-century discoveries of electromagnetism, was yet another impressive example of what science could do. So there began to develop a cult of the power of science. As time went on, this expressed itself in a general idea which is still with us: scientists can solve any problem if they will only set their minds to it! Almost simultaneously, the hope grew that scientists would make beneficial changes not only in realms allied to the hard core of science (that is, medicine, technology, agriculture, and so on), but also in human affairs. The nineteenth and twentieth centuries provide a long list of such hoped-for applications of science in the conduct of life, including "scientific" management, "scientific" sociology and—more recently—"operations analysis," a way of solving specific problems in World War II that seemed to many people to hold the promise of solving all human problems, whether housing, international relations, or the problems of the city.

* * *

The general appreciation of the tremendous potency of science in the realm of practical affairs, while justified to a very large degree by an impressive list of innovations, also had its own backlash. This manifested itself in the twentieth century as a fear of "technological unemployment," and was sometimes expressed in terms of an embarrassing fecundity of a science-based technology.

World War II provided an impressive documentation of the potency of a science-based technology, and this has been reinforced by the "early" successes of the space programs. Yet, at the same time, this great power of science has raised questions in many people's minds as to how the fruits of science should be used. Here I see a legacy from the nineteenth century of the fact that the scientist should not be trusted because he is tampering with nature. This was expressed in a very dramatic way in a short story by Hawthorne entitled "Dr. Heidegger's Experiment." I think it is notable that Dr. Heidegger was obviously either a German or of Germanic origin. Dr. Heidegger's wife was a lovely person in her character and was beautiful to look at, save that on her face she had a gross, disfiguring birthmark. As you may imagine, the doctor felt it necessary to make an experiment and to improve upon her beauty by removing the birthmark. He did so, and during the operation his wife died. Moral: Only death and destruction result when man tampers with nature or tries to improve on nature.

I need not remind you of the widespread feeling that has arisen against science in recent years as a result of man's tampering with nature, and the effect this has had on our environment. At this point, the scientist in defense has tried to argue that it is not his science that has destroyed the environment, but rather technology. The environmentalists and the antiscience forces reply that one cannot disengage science from the technology based upon it, and that scientists must bear the responsibility for the fruit of their ideas. The question here is not only the responsibility that an individual scientist may have for the uses to which his own work may be put, but also some responsibility of the knowledge system as a whole, the institutionalized system of scientific knowledge and power in which any given scientist is only an actor. It may be, therefore, that the responsibility for the applications of science lies in the system of knowledge rather than in the activities of a single person.

Freeman Dyson has just written: "It makes no sense to me to separate science from technology, technology from ethics, or ethics from religion." I am not sure all scientists would agree. I do not propose to go into the question here of the scientist's responsibility for the technologies based upon his discoveries, which would take me too far afield, although it is a topic directly related to the theme of this conference. However, I would like to know, and I must confess that I do not know, when it first began to be assumed that technological effects of science were to be blamed upon science itself. Alternatively, there is an interesting question as to when they became separated, since in the late nineteenth century it was generally understood by scientists and nonscientists alike that science was responsible for the material progress manifested in new technologies, new medicine, and new agriculture. And during the preceding two centuries, the first two centuries of modern science, it had always been understood that such practical applications would be the natural end-product of the new knowledge and would in fact justify the costs of doing science. Of course no scientist, considering Dyson's statement, could possibly doubt that today technology has become so science-based that it cannot be conceived apart from science,[11] but many scientists still try to cling to the ivory-tower view that whatever use is made of their science is the result of the will of the society at large and is not a primary concern of scientists themselves. I do not know whether such a disagreement with Dyson's brutally frank statement of his position would represent a case of special pleading or a carefully thought-out belief that there is an essential discontinuity between applied and basic or pure science.

During the Great Depression of 1929 and after, certain religious, political, and labor leaders called for a "moratorium on science," indicating their great fear that even more science would produce a greater technological unemployment. Although the ostensible basis for their position was concern for jobs, there is to my mind no doubt that these people also were expressing an antagonism and fear of science which, we have seen, was of long standing. They are saying in effect: there is an evil result for society that comes from tampering with nature, from not leaving things as they are.

It is often thought that this fear of tampering with nature is something which only nonscientists and people who are indeed antiscientific say about scientists. However, this is not entirely so. When the first atomic explosion occurred at Alamogordo, some of the scientists present, those who had been actively engaged in the productin of the bomb, thought only of apocalypse, of the end of the world; and there was even wonder as to whether man had not produced something by science that was beyond human control. Of course, the calculations had shown that this nuclear detonation would not start a reaction which would blow up the world, but the fact that at the moment of the explosion scientists themselves were concerned about the end of the world would seem to indicate that there is a fear and distrust of tampering with nature that even may affect scientists themselves.

I do not believe that the fear of science today, considered in its historic roots, is to be understood only in terms of practical issues such as the deterioration of the natural environment. Furthermore, I do not believe that the trust in science as the solver of all problems is to be attributed merely to the success that science has had in providing a basis for recent and current radical transformations of technology and of medicine and public health. Basically, there is a fear and a dislike of science because it seems like a juggernaut, something inhuman, something impersonal, something that affects the scientists themselves and makes them dehumanized and impersonal. This fear goes back to the notion of science as connected with the devil, with the forces of darkness rather than light, and with the Faustian bargain of which I have spoken earlier. It is this dehumanization which shows itself in the fear and distrust associated with Darwinian evolution and reductionist biology. It is this same fear that expresses itself in the antipathy to statistics, to the concept of reductionism, and to the diminishing of the stature of human beings by making them merely another set of numbers on punched cards in a data processing system.

Such fear and distrust is in counterpoint to the historical feeling that science is a force that extends man's ordinary powers. This line is associated with God's gift of reason to man, which he uses in studying nature, and—finding sermons in sticks and stones—celebrates the divine Creator by studying the world of nature that God has made. The wonder and creative force of science is coupled with the notion that this power given to man will enable him to do new things, and it is assumed that what is new is necessarily better, or at least tends to be better. Here the only complaint against the scientist is that he does not sufficiently use his knowledge and his power for the good of man. Indeed, there is a criticism implicit in much of the popular attitude toward science (and by popular, I include the whole gamut of nonscientists from government officials to the ordinary citizen) for not having used that enormous fecundity to solve such large-scale pressing problems as those of the inner city, the inhumanity of man to man, the ravaging effects of wars, and so on. Certainly the belief is widespread today that scientists "will" (and hence why should they not do so now?) find a new source of energy to replace oil. Of course, it might be answered that nuclear energy *has* been discovered and does just that, but there seems to be a general hope and belief that a more benign kind of energy will eventually be found.

There is one final aspect of the problem from a historical point of view that I have not explored. This hinges on what it is conceived a "scientist" is, or what "science" is. Academic people have one idea, but others have had a quite different one. There is even a difference here between the

German *Wissenschaft,* the French *science,* and the Anglo-American *science.* This is especially so with regard to all those important areas other than the traditional ones of astronomy, physics, chemistry, geology, botany, zoology, physiology, and so on. However, I think it is a fair observation to say, and this shall be my concluding note, that the area of fear and distrust is certainly as great—and in some ways is even greater —in the realm of the social and behavioral sciences (in which I would include applied genetics and human engineering) than in the natural or exact sciences. Furthermore, I would add that, from my own point of view, there is probably some real justification for this fear and distrust. Indeed, if the social and behavioral sciences ever were to achieve the power of the natural sciences, mankind might very well fear the consequences.[12]

NOTES

1. The research on which this paper is based has been supported by a grant from the Spencer Foundation. The twin themes of the conference were "fear of science" and "trust in science." My paper, however, deals chiefly with "fear of science" and "distrust of science," primarily because these twin themes seem to me to have been generally more significant during the long course of modern history than has been "trust in science."

2. In the discussion of my paper, there was some confusion with respect to the point I am making here. During the seventeenth and eighteenth centuries, scientists themselves did not as yet make the distinctions which arose only late in the nineteenth century, between "pure" science and "applied" science. The Royal Society of London in the seventeenth century was equally concerned with the possible production of a "history of the trades" (mining, metallurgy, boat building, agriculture, and so on) as with principles of biology and Newtonian physics. Two seventeenth-century scientists, Christian Huygens and Robert Hooke, designed the form of escapement which made the pendulum clock feasible. The *Encyclopédie* of Denis Diderot displayed side by side the main features of technology and the tools and principles of the sciences. During the eighteenth century, scientists influenced technological processes, notably in chemistry, where the effects of science were to be seen in metallurgical practice, the manufacture of gunpowder, pharmacy, and various other fields. Physicists and mathematicians were concerned with improving the science of exterior ballistics, and astronomers worked on problems of finding the longitude at sea. It would, accordingly, be wholly wrong to say that Franklin's invention of the lightning rod was the only example of a practical innovation based on science during the period from 1600 to 1850; it was rather the most dramatic. Additionally, the lightning rod was the only really dramatic practical innovation to derive from disinterested scientific research during this period. It must be kept in mind that the pattern with which we are familiar today, in which there is a sudden and far-reaching transformation of a whole area of technological or medical activity as the result of new scientific knowledge, is something which began to occur only since about the middle of the nineteenth century. It must be stressed that prior to that time, science was not the fount of innovation which radically changed the ways in

which men conducted their daily lives. It also had little, if any, direct effect on the general patterns of social interaction and exchange. Hence, the emergence of science as a really potent factor in the daily life of man, in the conduct of national and international affairs, and in determining the direction of the national economy and the national security, was not a feature of the scientific enterprise until only a century-and-a-quarter ago. Accordingly, the kind of fear and distrust of science which we have today as a result of the effects of a science-based technology were not fundamental aspects of the scientific scene during the two-and-a-half centuries following the emergence, in the age of Galileo, Kepler, and Harvey, of modern science as we know it today.

3. Of course, whoever was inoculated *did* get the smallpox, and hence it was correct to say that the practice of inoculation did spread the disease by giving it to individuals who might not have got it in the "natural" way. Furthermore, since there are many instances of individuals who died from the smallpox given them by inoculation (among them notably Jonathan Edwards), there was some valid ground to fear the new practice. Furthermore, the evidence in favor of inoculation was statistical, and we shall see below that there has always been a distrust of statistical evidence. Nevertheless, despite what one might guess to be the difficulty of determining whether or not a case of smallpox is "mild," and so should provide the source of matter to be used in inoculation, it is a matter of record that many physicians—such as Franklin's friend Jan Ingenhousz—performed many inoculations during the course of a long medical practice without any notable casualties. The popular reaction, in any case, was not based on specific scientific grounds so much as on a general uneasiness about tampering with the course of nature. The point cannot be overstressed in this connection that tampering with nature has special significance and gives rise to special fears and concerns when what is being tampered with is the human body. Perhaps the anxiety here is related to the notion that man is made in God's image and should therefore be inviolate. However, there is also a personal quality to such tampering in the case of the human body which has different psychological overtones from experiments with animals or plants or even attempts to control the forces of external nature.

4. In the *Encyclopédie*, science and technology were mixed, with no indication given that these might even possibly be wholly separate categories of human endeavor. No doubt, more people were attracted to the accounts of technological processes (the manufacture of gunpowder, the making of rouge and other cosmetics, military affairs, and practical pursuits such as those of the farrier, the cutler, the printer, the artist, and so on) than to the accounts of pure science, philosophy, or history. The illustrations in particular showing these practical arts or trades must have been particularly appealing to many people who either were not especially interested in the sciences or who lacked the ability to understand the at times technical accounts of physics, chemistry, biology and natural history, geology, and even higher mathematics. It is curious to observe, however, that no distinction was made between science for knowledge and science for utility during this period of the eighteenth century, nor had it been made during the seventeenth century, when science still had not produced great effects upon practical affairs. The current distinction between "pure" and "applied" science arose only in the nineteenth century, at a time when the utility that had always been considered part of "pure" science came to be of real significance.

5. Goethe was actually adding another dimension to what he considered the Newtonian way of explaining nature. He was aware of, and sympathetic with, notions of subjective sense perception rather than the mere objectifying of colors and the objectifying of light. From this point of view he is including as part of the epistemology of science the dimension of knowledge which he believed the Newtonians had not understood or at least not accepted.

6. It would be interesting to know the state of general public opinion in European countries with respect to Darwinian evolution. I have given two American examples only because they happen to have come within my ken. However, it is well known that Europeans have not accepted the Darwinian theory of evolution by natural selection to anything like the degree to which that theory has been accepted in England and in America. Even in England and America until fairly recently—that is, before what has been called the "Darwinian Synthesis"—many leading American scientists (notably geneticists and paleontologists) did not give full adherence to Darwin's notion of natural selection.

7. Additionally, the Darwinian system offered a challenge to some of the basic values, particularly religious values: a challenge to a firmly coherent alternate value system.

8. I remind the reader, as mentioned in an earlier note, that this statement is not meant to imply that there had been no earlier instances of scientists doing work that improved practice, or that there had been no scientific discoveries (other than the lightning rod) which had produced useful benefits. The point is rather that the kind of dramatic innovation in medicine, technology, and agriculture which is characteristic of science today was not a feature of the world of science and its applications until about a century-and-a-quarter ago. The reason that it is important to keep this historical fact in mind is that the attitude toward science as a potent source of innovation in our daily lives is not of as long a standing as the distrust or fear of science on more intimate psychological, intellectual, religious, or philosophical grounds, as has been discussed in my presentation thus far.

9. It is not without significance that the first large-scale dramatic example of the application of science to practical ends should have occurred in chemistry. The fact of the matter is that chemistry had long been the chief area of science which had found applications of a practical kind. For example, the production of acids and alkalis, the use of chemistry as an adjunct to metallurgy, the very fact that the production of drugs and of pharmaceuticals in general is a chemical process (although many of the traditional drugs were produced from plants or plant material) would indicate the ever-growing involvement of chemistry in the practical arts. Early in the nineteenth century, the practical effects of chemistry were widely debated in relation to the inorganic fertilizers in place of the traditional organic fertilizers and naturally produced nitrates. The chief figure in this new movement was the German chemist Liebig. The discussions about inorganic fertilizers took place just before the rise of the coal tar dye industry.

10. In this connection it is not without significance that the government support of the synthetic dye industry in Germany was not only based on industrial needs and the desire to capture the world market. The fact is that unstable dyes are explosives, so that while Germany was building up her synthetic dye industry, she was simultaneously constructing a potential source of manufacturing explosives for the supply of her armies.

11. In short, there is no question about technology being inseparable from science, although the issue is debated that science may be in some way separable from technology.

12. The danger here is that in the social and behavioral sciences, unlike the exact and natural sciences, there is ordinarily no separation between "pure" knowledge and its applications, since the situations being studied are those of practical life and social organization. Furthermore, many of the problems studied by social and behavioral scientists, or related to their studies, tend to be social problems of such pressing importance that there is an impatience for amelioration of such a magnitude that practical measures may be taken before it is abundantly clear that they are sound or even fruitful. There are enough examples of this sort to cause a real anxiety about the future.

Chapter 6

Discussion of
"The Fear and Distrust of
Science in Historical
Perspective"

Summarized by Andrei S. Markovits and Karl W. Deutsch

Hans Wolfgang Levi opened the discussion with praise for I. Bernard Cohen's metaphor of the lightning rod as a bridge between attitudes toward science in the past and today. The main difference between attitudes toward technology then and now, he believed, had to do with communications: negative attitudes toward technological innovations have greater significance because of the enormous power of the communications media. He also said that there is currently distrust of technology because it has been unable to find a substitute for oil; every attempt to do so, however, is seen as a new lightning rod.

Cohen interjected that he agreed strongly with Levi. He remembered his own attitude in a discussion with a colleague on the morning after the Three Mile Island accident. "In my innocence," he recalled, "I thought, how marvelous—even though something went wrong, there was still no catastrophe. How wrong we were!" Attitudes today, he tentatively concluded, are the same as they were in the nineteenth century. He confessed unsure knowledge, however, of what precisely the mass attitudes of hostility toward technology then were. He cited, for instance, the Luddites, or "machine-wreckers"; he also pointed out the difference between the attitudes of weavers and those of farmers cultivating the madder plant in the south of France, who had their livelihood altered by chemical advances in Manchester and Germany. The weavers only substituted a hand loom with a machine, but the farmers had to find a new crop. Cohen said that he was aware of the attitudes of certain literary figures, such as Charles Dickens, who was a great enemy of the railroads and of the field of statistics, which changed men to numbers; he handed over the question of mass attitudes to David Landes.

Landes commented that the labor movement, which would seem to have suffered most from technological displacement, was eventually won over to the notion that technology was a good thing. By the end of the nineteenth century, labor unions became partners in the campaign for technological advance; they wanted to control conditions of production, to protect workers and skills, but, with all that, the last thing they would say was that technology was intrinsically harmful. On the ideological side, of course, scientific socialism, Marxian socialism, was very pro-technology.

Peter Weingart posed a question concerning the nature of early fears derived from the lightning rod. A comparable modern development is the transistor and the technology arising from it, but the transistor does not pose the threat to ideology or to religion that people perceived the lightning rod to have done. Was the lightning rod seen as bringing catastrophe to mankind, he asked, or were its effects limited to those who used it? It seemed to Weingart that there is a fundamental difference between fears of catastrophe and fears of "incremental" harm.

Answering Weingart, Cohen said that, first, he did not mean to make so much of the lightning rod example; he meant only to give an instance of a technological application of science which, he said, served to show the hostility people hold toward bold new innovations. The slightest drought was blamed on the lightning rod because it interfered with the ordinary course of nature, its detractors said. They simply worried, like the people who believed that changes in the weather were caused by open-air atomic explosions, that one was upsetting the balance of nature and changing the environment, not just inviting localized destruction from a badly placed lightning rod.

As science began to gain its great power in the eighteenth century, Cohen explained, this became a constant fear. The implication of Hawthorne's "Dr. Heidegger's Experiment," as cited in Cohen's paper, was that if one tries to change nature, one invites death and destruction. This, he said, remains a very strong theme. The lightning rod is an important example because its perceived power to control nature, to change the environment, was greatly feared, and this kind of fear is persistent. And finally, he noted, such fear is not confined to nonscientists: at Alamogordo, "there was a feeling that, by God, maybe we're going to blow up the universe! They knew they shouldn't be worried, that the calculations had been made, but they worried anyway."

John Platt complemented Cohen's observation about the scientists at Alamogordo with a story. It is said, he related, that Fermi had calculated the probability that an atomic explosion would ignite the atmosphere and spread around the earth. He knew that it was a very tiny possibility, but on the morning of the explosion, "he went around trying to make bets of several hundred dollars with people, that if the bomb did not ignite the atmosphere, then *he* would collect; if it did, *they* would collect."

Taking up the theme of public distrust of interference with nature, Thomas Trautner argued that such a fear could be hypocritical. "People are continually playing around with nature," he said, pointing to breeding of domestic animals, the grafting of plants, and other agricultural developments. Farmers, he said, have been shown to have the least fear of genetic engineering of any social group, because in their work they are accustomed to alterations of nature. Thus, such a fear is not universal, but learned.

Horst Ohnsorge questioned whether it is indeed the great majority of the people that hold "the big fear." We know, he said, that this fear has existed, but perhaps it is not as strong today as it once was. Maybe it is the scientists themselves who distrust their own science and then inject that fear into open public discussion. Ohnsorge believed that people do not distrust science because they fear it cannot solve all their problems; to the contrary, "They believe that we can solve more problems than in fact we can." We should ask more often, he held, how science or technology can solve our problems. To inspire trust, we need to concentrate more on the positive aspects of science, not the negative ones that the symposium seemed to stress thus far. Fear and trust vary in cyclical trends over time, and at the moment, we should increase trust.

Karl Deutsch suggested that the question of whether it is indeed the masses, and not the scientific elites, who fear science, may provoke the public opinion experts at the symposium to tell more about these attitudes. Deutsch then treated the question of whether it is science that has fostered technology, or vice versa, as assumed in the contention that the steam engine did more for science than science did for the steam engine.

Deutsch suggested four instances of industries which have been called science-based. The first, architecture, is treated in an essay by George Santayana. In this it is argued that the dome of the cathedral in Florence was the first building constructed according to explicit mathematical calculations by its architect, Brunelleschi. Santayana holds that this achievement, circa 1480, marks the beginning of modern science, as it was the first time that man discovered that large masses of stone would behave exactly as mathematics said they would.

The second industry that may have been science-based was inter-continental navigation. "Navigators paid good money for research," Deutsch said, "on problems of how long it would take to get where." The very idea of sailing west to reach the East Indies was based on science: it was already down on paper somewhere, and at the point that the sailors boarded their ships, the search became an experimental endeavor of science.

The third was probably gunnery, which began as a craft. Somewhere around the sixteenth and seventeenth centuries, however, mathematics was used to calculate trajectories for artillery. The fourth science-based industry was insurance. It began with the shipowners getting together at Lloyd's of London, but by the time life insurance was offered, mortality figures were being compiled. Deutsch observed in jest that "the same churchmen who made contentious remarks about statistics probably bought life insurance." Though it is true that many of these mortality tables were inaccurate, "Science which makes no mistakes has not yet existed."

Deutsch also asked about the status of public attitudes toward forms of science that seem to increase the options of individuals, without particularly threatening them: long-playing records, transistor radios, other transistor technology, and, for the majority of people, television. Indeed, many members of the counterculture condemn science and, at the same time, are among the most promising customers of micro-technology in their use of pills that are popular "in quite a few circles of the counterculture." These are not considered a threat, but macro-technology, that which dwarfs the individual, evidently is.

Concluding his questions, Deutsch asked about the nature of electoral mechanisms with regard to science. He recalled that when his mother was a member of the parliament in Prague, representing the Sudeten-land, 3 or 4 percent of the population in her district opposed vaccination with religious intensity. They were a one-issue constituency and were capable of swinging an election to one side or another. Politicians were careful not to alienate this minority. "It occurs to me," he said, "that in the Federal Republic today, the 'green list' [environmental party] comprises 3 or 4 percent of the electorate, approximately the margin between the ruling party and the opposition, and I notice how friendly politicians are getting to that 3 or 4 percent." In a close election, such a minority becomes important; it may be that the communications media act as amplifiers for such small groups.

Cohen, in reply to Deutsch, noted that the question of vaccination was indeed very interesting, because vaccination seems to have spawned long-term resistance. One of the leaders of this resistance was Alfred Russel Wallace, the codiscoverer of evolution with Darwin, and he fought vaccination throughout his entire life. People opposed vaccination for a variety of reasons, but mainly because it interfered with nature's own selective processes, Cohen noted.

He then spoke to Deutsch's main body of comments, on the nature of science-based industries. Cohen first drew the distinction that, while Deutsch spoke of "science-based industries," he [Cohen] was more concerned with discoveries of new science which, *then*, in their applications, caused alterations in all these areas. The point about navigation, he said, was particularly interesting, because at the time of Newton, sailors were still determining their latitude by principles prescribed by Ptolemy. Not until the late eighteenth century were the methods of so-called lunar distances introduced, and even these never succeeded. The solution was only reached when John Harrison, a mechanical inventor with no scientific education whatsoever, made a clock that kept Greenwich time accurately and which, when compared to easily determinable local time, gave a ship's location within a few nautical miles.

Tables that were used for artillery also did not, in fact, rely on new science. Galileo found the parabola and thought that its use would improve gunnery, Cohen pointed out, but not only do shells not follow parabolas (because of wobbling), but barrels were so bad that mathematically-based accuracy was not relevant. Well-drawn mathematical explanations did not reflect the day-to-day practice of gunnery.

On the topic of architecture, Cohen said that, though Santayana did argue that Brunelleschi used mathematics to determine the shape of the dome of the cathedral in Florence, these methods were not new, and it is not clear whether Brunelleschi changed the way subsequent architects did their work. Cohen added to his evaluation of science-based industries, however, that pharmacy might be included, as might metallurgy. Even in these fields, he said, theoretical advances in chemistry probably did very little to alter actual modes of production.

Deutsch asked Cohen, "Do you then define science as essentially only that which follows a previously developed line of mathematical or deductive reasoning? Or would you include . . . also the systematic, or unsystematic, search for the unknown in nature?" Deutsch pointed to Böttger, an alchemist who started out to make gold but ended up inventing china. Would Cohen count that as science?

In reply, Cohen said that he would, but that he thought Deutsch was speaking rather of natural history—finding plants with medicinal value, for example. Cohen postulated that anyone working with new materials was in a sense engaged in scientific activity. One of his favorite examples concerns the Wright Brothers, who are considered simple mechanical engineers by many because they were bicycle mechanics. At a certain time, though, they discovered that they did not know what happened to wing structures when air was blown on them, so they built a wind tunnel and conducted experiments. At that point, Cohen held, they became scientists.

In response to Ohnsorge's question of whether he drew a distinction between science and invention, Cohen replied that he did, but that the distinction was difficult to define. An example, he said, that shows the complexity of the distinction is the telephone. When Maxwell, the world's leading authority on electromagnetism at the time, was asked to lecture on the topic of the telephone, he said that it was not of scientific interest, because it only used well-known principles of science in a simple way. In fact, Maxwell said, the only interesting thing about it was that "it talks." Even the great genius Maxwell, Cohen stressed, did not understand that in this operation there were being produced transient currents, and that this implied a whole new theory of which he had no idea. "So you see," Cohen concluded, "how hard it can be to make that distinction."

Cohen would nevertheless draw the line in obvious cases when someone makes a strictly mechanical linkage, improving a device like a reaper or a harvester or a sewing machine; he would not consider this science. He would draw the distinction rather between scientists and inventors, and this distinction rests on a difference in point of view. The inventor, if he finds something which is an interesting fact of nature, will ignore its theoretical implications because he is interested only in the practical. Edison, for example, discovered the "Edison effect" but ignored it because "he was not interested in the principles of nature at all." It had to be found again later by others who were studying discharges in vacuums.

Taking up the question of science-based industries, Landes said that he would differ from Cohen and from Arago; it was a surprise to Landes that Arago, coming from France, a country that perhaps more than any other had realized the potential link between science and practical application and had mobilized science in the service of the nation, could have been so unaware of the contributions made by science to technology. In 1831, when Arago made his statement, he could have chosen so many other examples than the lightning rod: the efficient production of explosives, beet sugar refining, and new processes for producing alkalis, all of which were discovered within his lifetime.

The invention of a precise timepiece, Landes elaborated, was another example of a symbiosis between science and technology. The first important innovation was Galileo's rough isochronism of the pendulum; when he died, he actually had plans for a pendulum-controlled clock, which was never completed. This work was taken up later by Christian Huygens's, who did some of the necessary thinking and went to a clockmaker (a craftsman), who then built the clock at Huygens's instruction. The next major invention, Landes noted, was the balance spring, which made the pendulum unnecessary, thus making possible the marine chronometer. Robert Hooke may have been the one responsible for this innovation, but what was important was the improved precision of timepieces that you could bring to sea, whose movements, unlike those of a pendulum, would be unaffected by the movement of the ocean. Clockmaking stimulated science by giving it an instrument it had never had before, and it benefited from science, which gave it principles of mechanics that the watchmakers and clockmakers did not know. The same symbiosis applies, Landes concluded, to the development of metal alloys later on.

With this Cohen strongly disagreed. Huygens's major scientific breakthrough, he said, was his idea that cycloidal pallets allowed wider swings of the pendulum and still kept perfect isochronism. However, in the pendulum clock, there are usually small swings; thus this principle was never used. Huygens was a good inventor because he had scientific training and a scientific mind, but his contributions to clockmaking were purely mechanical. He was acting principally as an inventor, as also was Hooke. The question of scientist-as-inventor, Cohen observed, has become more important since World War II, with the involvement of scientists in the military realm, with weapons evaluation groups and the like. There is no doubt that technology provided science with a great instrument for use in astronomy, for example. Cohen concluded, however, that Huygens and Hooke were acting only as mechanical inventors.

Landes defended his position, asking rhetorically, "These clockmakers were the finest craftsmen in the world at the time. Do you think it is coincidence that they got their inspiration from people like Galileo, Huygens, and Hooke? To me, that exceeds the bounds of probability." Mechanics could not have found the answers themselves, without the help of physicists, Landes reiterated.

Ohnsorge brought this part of the discussion to a close with his observation, in disagreement with Cohen, that "all knowledge belongs to science, and the inventor, by discovering new ways of doing things, adds to knowledge. Therefore, all inventions are part of science."

Max Kaase turned to two earlier comments by Deutsch. He first speculated on the extent to which we can learn from history whether alternating currents of fear and trust can be determined within that history, or whether such trends are indeed scholastic constructs; this would have a bearing on our predictions for the future. He then commented that the participants were probably not very clear on their reference group when they spoke of fear and trust, because it is likely that the general population had neither fear nor trust of institutionalized science, since without reading and without communication, there was little awareness. This distinction is revelant to the contemporary situation, he said, because we now deal with mass publics, influenced immediately by the results of scientific activity. Kaase concluded this point by expressing the hope that social scientists might reach a conditional insight on just who the carriers of fear and trust are, the public or the scientists.

In answer to another of Deutsch's questions, Kaase noted that little research had been done on the structure of attitudes and perceptions about science. Whatever evidence there is, he said, seems to show a fairly balanced view: people are still overwhelmingly in favor of science in the sense that it can improve their lives; on the other hand, this is balanced by an awareness of the negative consequences of scientific activity. This does not necessarily imply ambivalence. After a major science-related catastrophe, public attitudes toward science can have an immediate and major impact, in a democratic society, on the structure of science as an institution.

Kaase agreed with Deutsch that science as reflected in cultural criticism and skepticism can have an immediate effect on the electoral process under very specific circumstances. Referring to Peter Glotz's address delivered at the opening ceremony, Kaase noted that under certain circumstances, science can have a great impact on society in that a totalitarian or authoritarian form of government may have to be implemented because of certain developments.

Robert Lopez commented that he was particularly struck by Cohen's emphasis on novelty. The word *novelty*, he said, had been a "bad word" in the dictionary of Western civilization for ages; if there was truly a turning point, it was when the word no longer evoked such negative sentiments. Novelty, he believed, was first exalted when Girolamo Fracastoro wrote a hymn to the discovery of America, admitting thus "that the ancients did not know it all." However, Fracastoro was only one example; research should be initiated to determine when novelty is first seen as a good thing. Exactly when did civilization become more interested in novelty than in tradition? When was fear defeated by the desire to change for the better?

Lopez also commented that what is new about science today is "not only that it can kill everybody, which it could never do before," but also that our demographic problems bring us face-to-face with the realization that it may soon be physically impossible for all mankind to survive on the earth unless there is a change in the birth rate. Science is now in the position of being able to destroy mankind both by acting and by failing to act. Science has the greatest responsibility today; whereas technology was hitherto sufficient for problem solving, only science can respond to the dilemmas of our times.

Kaase responded that society is really in a better position than Lopez's comment would imply. Public attitudes, he said, are "really not that negative at all." Controlled, slow change is accepted by most people. It is only radical change that is strictly opposed: the public's concern about overtly radical change caused by science has increased, though this increase may perhaps be laid to political attitudes.

Responding to Deutsch's question as to what order of magnitude makes change qualitatively radical, Kaase said that the answer is not clear because none of the data he has seen have concerned themselves with the content of change. In the Eurobarometer data, there is an overall rejection, with variance among EEC countries, of radical changes in the political order. Germany underwent substantial change in attitudes over the last eight years, but the reason for that was strictly political; in 1969, the coalition between the Social Democratic and the Free Democratic parties came to power loaded with excessive expectations of change, and the disappointment following this promise resulted in a steady decrease in willingness to accept even gradual change. There is, however, a question of terminology when one discusses change.

Claus Offe then addressed a question to Cohen. Offe began by stating that the conference seemed to assume that there is a cyclical pattern between fear and trust in science; he wondered if it might be possible to discover the mechanisms of this cycle. It appears, he said, that the fear phases of the cycle are more extended historically, while trust seems to be an exceptional phenomenon. What are the conditions for trust? What specific motives incline people to trust science? There seem to be, he speculated, two groups of motives. First is the expectation that material wealth will increase, and that hardship and work will decrease, as a result of the subduing of nature. Second, moral and political questions could be solved by means of science: science could settle conflicts and wars, and has a unique capacity to resolve questions of moral philosophy, political economy, and political theory. These motivations seem to be quite diverse and heterogeneous. Is one of these motivations dominant in periods of trust? If not, to what extent do they mix?

Cohen answered that he believed Offe had raised a very important point. There do exist these two different strains in the trust of science: one holds that science provides a better and longer life; the other, that science solves large-scale social, political, and perhaps even moral questions. The second strain characterized the eighteenth century, when it was believed that one could view, through science, a natural order in the universe at large that could perhaps be applied to man.

But in the nineteenth century, Cohen indicated, the opposite view prevailed. Alfred Russel Wallace, in his autobiography, singled out the match—the ordinary phosphorous match—as the most important invention of the nineteenth century, precisely because of the practical nature of the item: man would no longer need coals to kindle a fire. Such inventions proliferated at the end of the century but were rare in its early years.

These two differing views of science and the scientist seem to merge in our time, Cohen suggested. One vision is of the absent-minded professor, a scientist lost in theory, who boils his watch while timing it with the egg. This vision is clearly inaccurate, since such a scientist would obviously never make much progress. The other vision is that of the preeminently practical scientist who applies all his findings to social uses. This is represented by the technocracy movement, as explicated by J. D. Bernal, who argued after World War II that operations analysis would solve all the world's problems: insufficient housing, underdevelopment, and so on.

Bruno Fritsch posed a final set of questions. "To what extent can we learn from history? To what extent are there really cycles of fear and trust?" Moreover, what did fear mean in the Middle Ages and in the eighteenth century, as compared to fear today? Elaborating on these ideas, he noted that we have to take certain thresholds into account: for instance, communication. In the past, ideology and religion hampered communication between scientists. There are no such restraints today, at least not to the extent that there were in the past. Another threshold is quantitative, with regard to the interaction between science and the environment. There was no interaction in times gone by; today there is a great deal. Perhaps, Fritsch suggested, trust and mistrust today are completely different phenomena than the same concepts exhibited in previous eras.

Cohen wished to comment on several points that had been raised over the course of the discussion. He stressed that the main point about the lightning rod example was that it presented to the public a visible symbol of a scientist's research. Without such a sign, a farmer might have never known that his scientist neighbor was engaged in research. However, if the farmer's crops failed, and he could see this lightning rod (which he did not fully understand), he created a connection in his mind; hatred and mistrust resulted. Another point that Cohen had been pondering concerned overt actions against scientists: persecution, destruction of laboratories, and so on. It occurred to him, however, that in most of the cases he could recall, scientists who had been persecuted and killed by mobs or governments had usually drawn hatred for their political beliefs, not for their research.

He then touched upon the question of science and industry, a problem which is central but which must have certain conditions attached to its discussion. We must distinguish, Cohen postulated, between very different kinds of processes. The first involves a scientific breakthrough—something new, like nuclear energy or coal tar dyes—which brings about great changes in technology. Compare this, he said, to smaller changes which may have great consequences in their mechanical applications. Huygens's work did not involve a major breakthrough in science; it was merely his great and wide scientific knowledge which, in its application, enabled him to reach a higher level of invention. The problem with scientists as inventors, then, was that scientists were usually unaware of what practical problems there were to be solved, as in Cohen's earlier allusion to the problem of Galileo and the pump.

Another kind of innovation, which came to the fore in the late nineteenth century, began in a very sudden and dramatic way to change people's lives and, therefore, to give them great hopes, Cohen said. The conquest of disease by scientists such as Louis Pasteur, for example, gave people expectations of what a science-based technology could do. These kinds of innovations and expectations, Cohen held, rest on a completely different level.

Cohen wished to add a few words about novelty. He agreed with Lopez that we should know when novelty became important, the point at which what was new was seen to be that which was better. In the eighteenth century, the age of progress, what was new and better was science. In the seventeenth century, Cohen said, Lopez's analogy to the New World was apt. Galileo's *Dialogues concerning Two New Sciences* and Kepler's *Astronomia Nova* were written then. Furthermore, when Ben Jonson wrote *News from the New World,* the new world was not America but the new world of the heavens, as discovered by the telescope.

The men who believed in progress liked this new knowledge, Cohen said, because it was secular and not sacred, and, above all, because it was democratic; anyone who followed the right methods could "do science." Descartes, in his *Discours de la methode,* claimed that he was never a particularly bright man but that he found a good method; hence he could make all these marvelous discoveries. No longer did a person have to be morally good and clean, or receive revelation from God to find truths; what one person did, anyone else could repeat. In religion, Cohen observed, beliefs had to be taken on faith, but this was not so in science. No longer did individual revelation prove truth; people could find truth and test it for themselves. Science was essentially democratic; it had reached every social level. Cohen maintained, however, that in the twentieth century, people came to see science as something the laymen could not understand. Legislators, in particular, could not hope to pass laws about science if they could not understand what scientists were doing. Modern attitudes harkened back to an older notion, demanding that science be intelligible to the public.

Cohen concluded by asking, "As science gets further and further from people, and more and more difficult to understand, where can they actually get hold of science? Where does the average person get his feeling for science?" Cohen maintained that the average person experiences science through its embodiment in technology. Though the layman may not be able to understand the functioning of a science-based instrument such as the laser, at least he can see it: he knows what it *does*. "It gives him an image," Cohen concluded, "of what science is and can do, what it might not do, what it ought to do. This is a real change, and should be considered as we look back historically: that the change is of such an order of magnitude that we don't have simple historical precedents to set paths for us to an easy understanding." With that the discussion ended.

Chapter 7

The Disenchantment
of Success

David S. Landes

I have a recurrent anxiety about addressing meetings like this one. I have a fear that I have prepared a talk that evokes no response from the other participants, and that, after a while, I come to realize that I must be speaking about something that has nothing to do with the theme of the meeting. In this instance, though, I am guilty of the inverse sin: I shall be covering ground that has already been explored by the two other speakers on this panel. No one told me, I am afraid, that I should confine myself to the relatively recent period of history, and so, you see, I shall begin my story with the Middle Ages and can only hope that both Professors Lopez and Cohen will forgive me for my trespasses.

But first a remark about some of the talks that we heard yesterday during the opening ceremony. Shepard Stone said, "I am glad Senator Glotz is here, because it takes money to engage in this kind of venture." Senator Glotz then called for (I quote) ". . . renouncing that which is possible but not desirable," hailed the abandonment of the supersonic aircraft by the United States, and called for SALT talks on technological development. He stated further that there was no gain in multiplying R & D programs and asserted that the application of new techniques should, in some cases, be stopped. In short, he recommended a policy of constraints and contraction. Instead of exchanging butter for guns, we would exchange ethics for both, and there is too much butter in the

European Economic Community (EEC) anyway. The result would be, of course, less money for Messrs. Deutsch and Stone, but Senator Glotz was surely not worrying about their requirements when he made his statement. (I do not want you to infer from these remarks an editorial position because, as the French say, *Je constate,* "I am just making an observation.")

Some more preliminary remarks: I read in this summer's newsletter from the American Automobile Association, which is a pressure group for American drivers of automobiles, the following headline: "Man on Moon Effort Needed to Solve Our Energy Problems." Another headline, this time in *The New York Times:* "Dalai Lama Says Buddhists Can Learn from Christians' Activism." (This was issued from a suite on the twenty-third floor of the Waldorf-Astoria Hotel in New York, which is a manmade version of Tibetan altitude.) Also from the *The New York Times*: "U.S. Urged to Relax Curb on DNA Studies." (This came from an advisory group to the Director of the National Institutes of Health of the United States.) In the same paper: "Plea by the Director of the National Academy of Sciences for Continued Government Support of Pure Science." He cited, among other things, some research on the densities of insect populations. It may seem silly to count insects, he said, but in fact this will provide very valuable information on the efficacity of various methods for reducing overpopulation of humans— surely a concept which will horrify some religious groups, which do not like to think of people in the same way in which they think of bugs.

As you can see, the sides are drawn. Whatever we may say here about fear and mistrust of science, there are obviously those who want to continue pushing towards scientific and technological advance. Indeed, I do not think it unreasonable to say that *most* people still have considerable faith in science, faith sometimes alloyed with anxiety. My sense is that man, especially Western man, really knows no other way to solve his problems.

With these thoughts out of the way, I want to turn now to the historical experience of this problem. But first, let me propose a definition. I know this is a dangerous thing to do: so far everybody has assiduously avoided defining science. One speaker evaded the issue by saying that you could count any new knowledge as science. That is too broad for me. For the purposes of my discussion, I will define science as the systematic pursuit of knowledge by human means, that is, by perception and thought. First comes the observation, usually deliberate, of things and events; then reasoning is employed to interpret and understand these observations (perceptions) and generalize about them. Such reasoning defines the objects of observation, that is, reality, in a dichotomous way: it accepts as real only those things and events that can be perceived, apprehended, and described by two or more people in the same terms. Such things and events fall in the realm of the natural, and all explanations of them must fall in the same closed set; that is, they must be apprehendable and verifiable by two or more people in the same terms. All supernatural phenomena and explanations, that is, things subjectively felt or intuited, are by definition excluded from the scientific domain.

Historically this distinction has great importance. The dichotomy between natural and supernatural implies an invidious distinction between science and other forms of knowing. Earlier in our discussion, someone argued that scientific knowledge is knowledge that cannot be known by everybody, but rather by only a small group with special training and with the intelligence to use it. He defined science, in other words, in terms of its limited constituency. This is surely true, but supernatural knowledge is also characterized by a limited constituency. What really matters here, I submit, is the nature of scientific matter, the manner of its apprehension and collection, and the mode of analysis and explanation. The point is that throughout history there have been other modes of learning and explaining—often modes dominant politically, culturally, and spiritually—and these have naturally perceived science as a threat to their preeminence. Hence much of the historical record of the fear and mistrust of science has been the response of vested interests to a dangerous challenge.

Now you all know that science in the sense that I have used is not very new. It goes back to ancient times. However, as a widespread, dominant mode, pretending to a monopoly of real knowledge, it goes back essentially only to the seventeenth century—to those intellectual innovations that we have often called the Scientific Revolution. The success of science in that period and subsequently was based on two things—first, the ability to learn, that is, to create new knowledge; and second, the ability to use that knowledge to do things that yielded material improvement and gain. From the start, the prestige of science, and with it the support it has received from political and economic elites, has been linked to its material applications. Because of this, I think our discussion must include the fear and mistrust of technology along with the fear and mistrust of science.

A number of speakers at the conference have argued that science-based technology is a relative newcomer, that it is not really until our own century that pure science has consistently led the way and nurtured a continuing flow of technological change. There is much truth in this, as any student of the Industrial Revolution knows. Most of the major innovations that made possible the breakthrough to the modern industrial era owed little if anything to pure science. They were rather the work of practical empiricists who used a mix of experience, observation, and common sense to imagine new ways of working and new devices to employ. In all fairness, though, this so-called common sense was not unrelated to larger changes in observation and reasoning promoted by the new science of Galileo, Newton, and their successors. As James Watt, the inventor of the improved steam engine put it, he did not get the idea of the separate condenser from Professor Black, the chemist, and his theory of latent heat, but he did learn ways of thinking about natural phenomena.

Any depreciation of the contribution of science to technology before the contemporary era, in the eighteenth century, for example, is justified only in terms of what has happened since. Observers of these matters going back to the Middle Ages were always aware of the contribution, actual and potential, that science could make to technology. One has only to examine the record of the support that scientists received over the centuries from ambitious rulers who saw them as one more source of power. This process culminated during the French Revolution, when a desperate government, forced to defend itself against a European coalition and concerned about its ability to obtain the raw materials of industry, mobilized the greatest scientific minds in the country and undertook a national effort to replace trade by brains and create substitutes for traditional natural resources. (Keep in mind that this was over one hundred years before Germany made a similar effort in World War I.)

In other words, whatever reservations contemporary historians of science may have concerning the contribution of science to technology in earlier centuries, clearly rulers and workers of those times thought differently. As a student of economic history, I agree with them. The neat divisions that we make today among theoretical science, applied science, and technique would have made no sense at all to earlier generations, who vaguely conceived of all of these as steps along a thoroughly blurred continuum. Scientists (at least those we think of as scientists) in those days were craftsmen as well as thinkers; they made their own tools and machines, and they ground and polished their own lenses. Their scientific thinking was often inspired and stimulated by their search for solutions to practical problems, while their scientific speculation suggested applications of great moment (cf. Galileo, the pendulum, and the pendulum clock).

In all this, I do not want to depreciate the ingenuity and the remarkable contribution of empiricist craft skills. I would be the last one to do so, as my description of the process of technical change during the Industrial Revolution shows. But neither do I want to sin in the other direction. We must not underestimate the growing importance of the contributions from the side of scientific speculation and research.

Nothing illustrates this better than the improvements in navigational science from the fifteenth through the eighteenth century. I am reminded here of a scene in *Treasure Island,* the novel of Robert Louis Stevenson, in which the pirates, who are getting ready to take over the ship, debate among themselves when to kill off the captain and the mate. You will remember that at this point Long John Silver persuades the others to wait until they have started back home. He reminds them that it would be unsafe to do otherwise; that although they know how to sail a ship, it is a very different thing to set a course. Only the officers can do that, because they have the training in astronomy and mathematics to do the necessary observations and calculations.

So much for the alleged lateness of science's contribution to technology. Returning now to our theme, that is, mistrust and fear of science, I want to start by distinguishing between two component aspects: (1) science as generator of knowledge; and (2) science as generator of technique. Let me follow these in turn.

For a long time, scientific knowledge as we understand it seemed to inspire neither fear nor mistrust. In the Middle Ages, for example, some clerics and scholars speculated about things that we would define as scientific. They posed no problem to the authorities or to the society as a whole, though, because they dealt largely with esoteric matters and did not as yet call into question any of the fundamental principles and values of the time.

Insofar as these speculations produced technical applications, they seem once again not to have engendered misgivings. Some of them, navigational science for example, were clearly perceived as socially desirable. The one innovation that might have given rise to the most serious anxiety, the application of gunpowder to warfare, seems hardly to have ruffled the security of contemporary observers. One or two deplored the destructive uses of gunpowder, but the great majority were happy to seize on it as an instrument of self-aggrandizement or self-defense.

Clearly, though, such serenity could not last forever. The potential for social conflict and fear was there from the start if only because European technology had the salient characteristic of being labor-saving, hence job-eliminating. In the Middle Ages, for example, the great innovations were water mills, windmills, the substitution of the horse for the ox in agriculture, metal-shaping and wire-drawing machines, and so on. All of these were things that substituted inanimate devices for human strength and skill, and this substitution obviously carried with it the threat of competition between new ways and old, with attendant risks of unemployment.

To be sure, in the Middle Ages, these dangers were not acute. From the millennium to about the fourteenth century, there was a long expansion, marked by an increasing demand for labor, which was then followed by a sharp contraction triggered by the Black Death (1348–1351). This contraction, for all its severity, was uneven in space and time. There were places that throve and others that faltered. The point is that, insofar as there was a contraction, it was accompanied by or brought about a sharp reduction in population, so that there was on balance a continuing labor shortage. Labor-saving devices, then, did not yet pose the kind of problem that they would present later on.

The question still remains as to why Western Europe was so favorably disposed to this energetic pursuit of new knowledge and new ways of working. This issue is very different from the one we have been discussing, that is, why Europe was not disposed to fear and combat these things. It is a very important question, because it is apparent to the student of comparative history that Europe was unique in this regard. Indeed, on one level, it is this pursuit, for better or worse, of scientific knowledge and technological advance, that sets Europe on a different path in the Middle Ages and eventually enables it to dominate the rest of the world. I repeat: for better or worse. I will not attempt to judge whether this has been good or bad for mankind; I am simply making an observation.

To be sure, European civilization in this regard is not a single, homogeneous set of values. Other traditions, other values, exist alongside. In particular, there is a very strong stream, going back to ancient times, which is profoundly suspicious of scientific speculation, of new knowledge, of tampering with nature, of changing what is and, hence, what ought to be. We have always feared the ambition that drives man to make himself the equal of the gods. We find this fear in the Jewish tradition (the story of the Tree of Knowledge); we find it in the Greek tradition (the stories of Icarus and Prometheus). It continues into the Christian era: the Church has always concerned itself with the dangerous possibility that man may come to worship himself (the Pelagian heresy).

This older tradition of fear and of hostility to change was, if anything, stimulated and exacerbated over time by the successes and the consequent prestige of the cult of change. Even so, it lost ground while the worship of newness rose *crescendo* to some kind of peak in the nineteenth century. Nowhere else in the world has there been so much admiration of new things for newness' sake; nowhere else has novelty taken on such positive connotations; no other society has accorded so much respect, admiration, and imitation to youth.

Now if you ask why this distinctive Western worship of change and "progress" occurred, I can only say that the answers are still being sought. This is clearly one of the great problems of world history; yet for all the work that has been done, there is still room for a great deal of further study. Let me quickly review, though, some of the suggestions and hypotheses that have been put forth.

Some writers have stressed the Judaic-Christian tradition, the biblical injunction to man to master the earth (*Genesis,* I, 26; IX, 1–3), which is contrasted with the widespread animist notion of the divinity of the nature around us—every stream has its naiad; every tree its dryad. If you believe, as most people on earth have, that there is something sacred in all the things around us, it becomes very hard to treat these things with a sense of dominion and the free right of use and disposal.

Other writers have pointed to the activism of Western Christianity, by contrast with other religions including Eastern Christianity (the Martha-Mary antithesis—Martha, activist; Mary, contemplative). Here I would recall to you the wistful words of the Dalai Lama cited earlier, to the effect that maybe we (the Tibetans) could have done with a little more of the activism and less of the contemplation.

Returning to a point that I made earlier, I would emphasize the fact that the West has developed a special, restrictive notion of reality. Reality is in effect defined as "What you see, I see." If you look at a thing and see a flower and I look at it and see a flower, we know that the flower is real. Anything beyond that is in the realm of the supernatural, of the fantastic. This conception differs radically from that of civilizations and cultures that have argued that everything, subjective and objective, is real, and that have devoted much of their talent to erasing the barriers between what appears to us in the West as two domains. This separation, I may note, is inculcated in our children very early by a wide range of devices, including fantasy literature that carefully distinguishes between real and unreal. It is at the heart of our whole process of education, and it lies at the basis of scientific inquiry as we know it, because it is this definition of reality that makes for the power of scientific observation. What you see, I see; and when I tell you about it, you will know what I am talking about. But if there is something that I just see within me, in my heart or soul, if I "imagine it," dream it, feel it, there is no way I can communicate it to you in replicable, verifiable terms.

Other explanations have emphasized material rather than, or alongside, intellectual factors. Thus the West has been favored geographically, in its climate, rainfall, and soil; and these happy physical circumstances have combined with a relatively high land-to-man ratio to make possible higher production per head and eventually the release of a greater proportion of society for nonagricultural activities.

Political circumstances have also made a big difference. By comparison with the other great civilizations of the world, European society has been characterized by a high degree of political fragmentation and interstate competition. It has consisted of a fairly large number of independent units striving with one another for place and power; hence, it has been very much concerned with the sources of power. It is of great moment that, very early on, European rulers perceived knowledge as a source of power, and it is no accident that these same rulers competed with one another to attract men of science, subsidized scientists' research and work, and encouraged them to apply their findings to practical purposes. This intellectual competition and emulation, with or without political support, generated a multiplicity of points of invention and innovation. Each person could build on the achievements of his predecessors, gain followed gain, and the result was a cumulative process of learning and doing. In this regard, Europe again differs from other parts of the world.

Let me give two historical examples of what I mean, chosen from a comparison between China and Europe. Why China? The answer is that, if one had to pick a single place where one would have thought that the advances that occurred in Europe should have occurred even earlier, China was the place. I remind you in this regard that when Marco Polo went there in the late thirteenth century, he was extraordinarily impressed by the superiority of the economic system and the ingenuity and skill of Chinese craftsmen and technicians—and Polo, after all, came from the richest and most advanced city in Christendom, namely, Venice. Yet China does not maintain its superiority. It reaches a kind of peak and then retreats. The greatest achievements of Chinese technology, for example, remain isolated *tours de force* rather than steps to new and better accomplishments. Meanwhile the West continues to move ahead, and soon, certainly by the fifteenth century, the advantage lies the other way.

Two major innovations will show what I mean. The first is the mechanical clock. The mechanical clock, weight-driven and with escapement regulation, was invented in Western Europe in the late thirteenth century and remained an effective European monopoly for 500 years. In Europe, its importance grew steadily, and it became in the long run the principal means of giving order to life and work. From the sixteenth century on, the Europeans began to export their timepieces to other civilizations, such as China. These civilizations in their turn placed a very high value on these devices, for which they had no equivalent. Even so, they never learned to make anything comparable themselves until the contemporary era; and what is more important, they never really knew what to do with these instruments. For them, the clock was a highly ingenious toy, more interesting for its automata and bells than for the time it showed on its dial.

Now I submit that this is surely not an accident. We are talking here about a device which is closely linked to the very character of our civilization. The Europeans learned to make ever better timepieces, because time measurement was, from the Middle Ages on, of the utmost importance to them.

What about the Chinese? Well, if you read the book *Heavenly Clockwork* by Joseph Needham, Wang Ling, and Derek de Solla Price, which is our best source on the subject, you will see that in China clocks were a monopoly of the Imperial Court. The clocks of the emperor were sometimes extraordinarily complex and ingenious instruments; however, they were not intended to measure time, but rather to imitate the movements of heavenly bodies. They were designed to facilitate astrological and calendrical calculations. These were crucially significant in a society where the emperor gave the signal for sowing and reaping and was guided by the stars in the highest and lowest decisions of the realm. There was in China no multiplicity of technicians and mechanics trying to find better ways to mark and keep time. Instead, *ad hoc* teams were gathered from time to time to build new mechanisms, each of which tended to fall into disuse and eventually disappeared after the death of their creators. As Needham says, over a thousand years of Chinese horological history can be summed up in the achievements of four or five men. No better indication could be given of the difference between China and Europe. There can be no diffusion and accumulation of technique when four or five men produce as many peak performances over a thousand years.

Gunpowder offers a somewhat similar story. In this instance, the invention is Chinese. Not until some hundreds of years had passed did Europe learn to make explosives. The Chinese used gunpowder both for peace (fireworks) and war. They never made the most of its potential, however. The Europeans leapt on this device when they learned about it, improved on its composition and, by preparing it in small corns or pellets, obtained substantially more rapid and violent explosive reactions. In addition, they had a large pool of excellent bellfounders, who had learned their trade by making church bells and could now shift over to the manufacture of cannon. By the time the Europeans returned east in the sixteenth century, they brought with them guns far more potent than anything the Chinese or, for that matter, the Indians, knew.

To sum up, we are talking here not about differences in intelligence, but in style and values. Invention and innovation were, from a very early period, much sought after and highly valued by Europeans. Whatever the force of an older antipromethean tradition, it was not strong enough seriously to inhibit this newer pursuit of the bigger, better, faster, and stronger.

The resistance to novelty does not gather strength until the so-called modern period (1500–1800). These centuries see the opening of the world to transoceanic navigation, the transformation of science, and the Industrial Revolution. Now, for the first time, science as a source of knowledge posed a challenge, explicit and implicit, to established moral and spiritual doctrine. This was no small matter. The impact of the threat varied from one part of Europe to another. There were Protestants who were unhappy with these new ideas; likewise, there were Catholics. On balance, though, the Protestants found it easier to live with them. Why? Largely because Catholicism recognized a single source of spiritual and intellectual authority and found it impossible, at least at first, to accept intellectual autonomy. This revulsion against intellectual innovation was aggravated by the Church's hostility to religious dissent. Sensitization to one kind of discord easily diffused into a general allergy against any and all kinds of intellectual iconoclasm. The Church of the Counter-Reformation was a very defensive, proscriptive, and prescriptive institution. It had little patience for heretics and geniuses.

By way of contrast, Protestant Europe, which was not necessarily more tolerant of disagreement, was compelled to live with it by its political fragmentation. Religious and intellectual dissenters quickly found that if life was impossible in one polity, they could remove to another. The more liberal Protestant states—Holland is the best example—became international havens for Europe's intellectual, spiritual, and economic rebels, so that with time, the same force for invention and change that separated Europe from the rest of the world began to separate northern from southern Europe.

Not until the nineteenth century did Protestant Europe begin to have trouble with science as a generator of knowledge. The difficulty came with the introduction of new ideas concerning the age of the Earth and the evolution of man. Another irritant was Bible criticism, which put itself forward as a science and called into question the scriptural bases of religious belief. Yet such resentments as did arise in Protestant areas as a result of these implicit contradictions of scripture generated more amusement than action. By this time society had become much too secularized, much too incredulous to take these defenses of religion seriously. Those places that did—for example, the hills of Tennessee— became an object of general risibility.

More serious in this more recent period was a growing concern for the direct and indirect effects of science and technology on economic welfare. The increasing pace of technical change, which picked up significantly in the eighteenth century, emphasized the demonic (to use Fritz Redlich's term) as well as the constructive aspects of the market economy and free enterprise. Older branches of manufacture were bypassed and left to stagnate and die, at great cost to workers unable or unwilling to learn new ways of working and accept new conditions of employment. At the same time, the newer branches were even more ruthless in the conditions imposed on their work force. In all fairness, these conditions varied considerably from place to place and from employer to employer. Even so, the growing concentration of pale, often deformed and badly nourished proletarians frightened contemporaries, who described this condition as a new kind of white slavery. The result was not only a factory reform movement, but often a nostalgic hankering after a mythically simple and happy past before the coming of the wheel, the steam engine, and the clock.

It would be a mistake, though, to overemphasize this revulsion against the sins and evils of a new machine age to the point of ignoring the growing prestige of science and technology among the population at large. For that matter, even those who might be expected to be most alert to the negative aspects of the changes that were taking place, economists, whether radical or conservative, were unwilling to condemn the knowledge that had made it all possible and promised still better things ahead. The socialists, for example, whether scientific (that is, Marxian) or utopian, refused to reject science and technological change *per se*. It was the use that man made of his knowledge, the system in which he gained and applied it, that was responsible for misuses and abuses.

This was the radical, critical position. The more conservative elements in an increasing affluent society took even greater satisfaction in the changes around them. Artists painted the early factories as romantic additions to a bucolic landscape. We have drawings of foundries that resemble nothing so much as the forge of Vulcan, modern evocations of classic mythology. For every William Blake, wounded in his deepest being by the uncompromising mechanism and rationality of the new industrial era, there was a satisfied and self-congratulatory Macaulay to sing its praises. The bulk of the population saw science as the key to a wonderful future. In this vision of linear progress, the Great Exhibition in London of 1851 was a celebration of material achievement, above all of British achievement. This example was quickly picked up by Paris and other capitals; so that the nineteenth century became a succession of commemorations and celebrations throughout the Western world.

Note that the very intellectual character of science lends itself to such a sense of progress: every day we know more and can do more. Inevitably, this entails a high rate of discount on past accomplishments. The scientists of today do not have to read those of even ten years ago to learn their subject, much less the works of earlier predecessors. Contrast in this respect history, whose practitioners can always learn by reading those who went before. History is clearly not a science in the same sense as are physics or chemistry, for example. Scholars in all these disciplines climb by standing on the shoulders of their predecessors and seeing farther. The natural scientists, however, having climbed, do not have to look behind them.

This confident, even complacent, mood changed in the twentieth century, which has been described as an age of disenchantment. A primary reason was World War I, a shattering revelation that man was not necessarily getting more reasonable and more civilized. It was a bloodletting, on a scale never imagined, and it irreparably destroyed the old political order. It left Europe exhausted, materially and morally, and called into question the legitimacy of European political leadership and imperial dominion.

The war was followed by Fascism, then world depression, again on an unprecedented scale. Now the economic order seemed to be crumbling, and if radicals and revolutionaries could take comfort in the darkness before the dawn, the middle class seemed to lose all courage and retreat into a hapless, ostrichlike conservatism. Only the predators and bullies seemed to be thriving.

By comparison, World War II was actually something of an upbeat—good triumphing over evil. Then came the postwar period, with its euphoric sense of catching up, of doing all the things that people had long wanted to do, of buying all the things that people had once been too poor to buy. In Europe particularly, this was a very bullish period—years of autos, refrigerators, modern plumbing, and central heating. Now it was the turn of the radicals to be disappointed. The Communist (Socialist) god had failed. Even the most tenacious will not to know could not indefinitely resist the evidence of tyranny and oppression in the non-capitalist countries. There was a brief moment of hope in salvation from the Third World, but this too was soon disappointed by evidence of brutality and inhumanity on a scale that surpassed all but the worst manifestations of European barbarism (thus Uganda and Cambodia). It has been hard in our time to retain our belief in linear progress. Only a religious, antiscientific faith could ignore the lessons of the evidence.

Our century has also been a time when science has become more and more esoteric and anti-common sense; the scientists, a new kind of priesthood. As a result, it has become harder and harder for the general population to identify with them and their work. There was a time, say, in the eighteenth century, when an educated man would have felt an obligation to learn and to know science and would have thought of science as part of his larger domain of interests and competence. People like Goethe, though primarily writers and poets, studied science seriously and were not afraid to write about it. Not everyone can be a Goethe, but those who contented themselves with learning and consuming nevertheless thought of scientists as part of a larger army of progress in which they (the consumers) shared.

By the twentieth century, this had become impossible. Too much science was now based on mathematics (in itself an arcane language) and dealt with matters not apprehendable by unaided sense perception. I remember from my own childhood an extreme example of the always arcane, hence, hieratic character of scientific knowledge. When I was a boy there was a man named Einstein, the word was, with a theory that only three or four people in the whole world could understand. He was the new high priest: around him were a handful of qualified assistants; from them, knowledge and competence trailed outward and downward to people like me.

Now this story may sound silly, a typical example of popular mythology. Nevertheless, like much popular mythology, it is not completely divorced from the truth. I have been told that in the Institute for Advanced Study at Princeton, the pure mathematicians look down on the physicists; the physicists look down on the other scientists; and all the scientists together look down on the social scientists. When Oppenheimer was the director of the Institute, he was perceived as a has-been—someone who was no longer up-to-date. Oppenheimer in turn got his consolation by reading Sanskrit and thereby affirming a compensating breadth of humanistic interest.

So it is with all hierarchies. If the scientists themselves draw lines, who can blame lesser people from doing the same? And lines mean alienation. Small wonder that there is a certain *Schadenfreude* throughout the intellectual community when scientists come a cropper. Remember the welcome given to Heisenberg's uncertainty principle (1927) and its hasty generalization to a wide range of knowledge to which it is simply not applicable. What one had there, I submit, was an unconscious revolt against the smugness and complacency of positivistic science. Sometimes the very people who most wanted to imitate science were the quickest to seize on these evidences of its limitations. Imitation may be the sincerest form of flattery, but flatterers generally do not like their models.

The most energetic of these flatterers have been the so-called social scientists—the very name is evidence of the pretension. If the scientific method could be so effective in opening the secrets of nature, why could not the same method be applied to a closer and truer study of man? The difficulty, of course, is that the social sciences must deal with far more complex phenomena than the natural sciences. The matter of study is less amenable to simplification and abstraction, hence less reducible to those mathematical forms of expression that we think of as "laws." To put it differently, reality here is not the same as in the natural sciences. We are no longer in the realm of "what you see, I see." In the social sciences, a person looks at a glass and says that it is half full; someone else looks at it and says that it is half empty. That is what social science is all about—a matter of more or less educated opinion. Scholars can often look at the same data and come away with diametrically opposite views.

The difference, then, between the social and the natural sciences would seem to be obvious. However, insofar as the social sciences have successfully asserted a claim to a share in the prestige of science and used this claim to justify a growing influence on social policy, their wisdom and the measures they have inspired have generated in many quarters hostility, opposition, and even doubt and scorn. Furthermore, by a kind of inferential contagion, this hostility and skepticism has communicated itself to the natural sciences as well. In that regard I find it interesting that one of the papers at this meeting counsels a general intellectual skepticism and tells us not to let even mathematical proofs compel our assent. The argument is, if you do not understand what you are reading and where it came from, you should not be compelled by it. On one level, this sounds reasonable. On another, of course, it completely undermines the intellectual authority of such esoteric disciplines as the natural sciences. Maybe we have put too much faith in experts; but can we afford to live without them?

The growing ignorance of science and the concomitant alienation from scientists have been reinforced by an increased disapproval of the works of science. As noted above, this is hardly new; there is an age-old tradition that condemns man's *hubris* in pretending to be a god. The nineteenth-century story of Dr. Frankenstein and his monster is a modern expression, with electrical props, of the *golem* myth; its suggestion of the scientist as fomenter of disaster has been picked up by a legion of other writers since. Now the scientific and technological advances of the twentieth century have given credibility to a literature that was once fantasy. For the first time we can contemplate the real possibility of the destruction of the world, of the creation of dangerous mutants, of ecological catastrophe, and the like. The result is that we now have meetings like this one and papers like those of a number of our contributors, full of fear and caution, even among the scientists. If the high priests are worried, what is the ordinary layman to think? If the scientists are afraid that they have become sorcerers' apprentices, we laymen ought to be terrified.

A final paradox: the very success of our science and technology has fostered a sense of alienation and embarrassment. What is it, after all, that more than anything has set us apart from the rest of the world, given us the lion's share of its riches, given us dominion over those less advanced in knowledge and technique? It has even been argued that this disparity between fat and lean, rich and poor, is not simply our reward for knowing and doing, but a penalty that we have inflicted on the rest of the world; that our development has been won by "underdeveloping" them. (The most vigorous exponent of these views has been André Gunder Frank, who argues primarily from the economic and political experience of Latin America).

In itself such an outcome should not discredit science, as any good Marxist can tell you. Yet in a strange way it has, by promoting a sense of guilt and encouraging the notion that science, and with it, technology, have gotten out of hand. The servants, it seems to me, have become the masters and turned us into instruments of destruction and exploitation. I remember in this regard a talk that Herbert Marcuse gave at Harvard fifteen years ago. He was full of despair. The advanced nations of the world had been hopelessly corrupted, he said, in values, personality, and institutions, by their worship and pursuit of technology. Our only hope lay in the nations of the Third World, still innocent because they had not yet eaten of the Tree of Knowledge. It was a strangely religious expression of faith from so unreligious a man, and nothing that has happened since would give him comfort in his conviction. On the contrary, while our misgivings have multiplied, the commitment to science and technology has if anything grown stronger in the Third World, which brings me to my last point.

It is a great historical irony that this growing fear of science comes at such an awkward time in history. To be sure, if these fears are justified, they cannot find expression and evoke action too soon. We had better start worrying more about what we are doing, because some of these things we are doing may well be irreversible.

Yet I say that this fear comes at an awkward time because it is just now, in our era, that the great majority of the people of the world have become conscious of the fact that they have suffered grievously for their want of scientific knowledge and technical advance and have determined that they must do something to make up the gap. In effect, we who have led the way now have the whole world on our heels. They want to move ahead along the same lines we have. They want more science, more technology. We may have begun to have reservations, to call for caution, to seek another way. They—the poor, the hungry, the uneducated—are impatient to leap ahead and are inclined if anything to view our cautions as subtle or disguised deterrents to competition from below. Having climbed the ladder, we want to pull it up after us.

Now it is hard enough to give effect to legitimate fears and concerns regarding science and technology in advanced affluent societies. It will be next to impossible to impose these precautions, which often entail substantial outlays, on the developing countries of the world. In short, however desirable the constraints that Mr. Glotz advocated, it will take more than our consent and determination alone to put them into effect. Moreover, the negative consequences of failure here will be far greater than those for which we have been responsible. We are talking here of the effort of three-quarters or more of the world's population to make up a substantial gap in material well-being and income per head. If you think that we, the advanced countries, are guilty of polluting, wait until you see what the whole globe can do.

In conclusion, let me restate this ironic paradox. Thanks to science and technology, a minority of the world's population enjoys a standard of living higher than even the kings of the past would have thought possible. Yet within this same minority are those who most acutely mistrust and fear the actual and potential achievements of science and technology. Meanwhile, on the outside of the window looking in, with noses pressed against the pane, stands the great majority of the world's people. They want an equal share of the goodies and have no qualms about the science and technology that they feel can help them get it.

It is not easy to think of a solution to both concerns and aspirations.

Chapter 8

Discussion of "The Disenchantment of Success"

Summarized by Andrei S. Markovits and Karl W. Deutsch

John Platt began the discussion by speculating on the role psychopharmacology played in Western expansionism and by asking what research, if any, had been done on this fascinating subject. In the fourteenth, fifteenth, and sixteenth centuries, he commented, Europeans were introduced to tea, coffee, chocolate, tobacco, and distilled alcohol spirits. He noted the importance of coffee to the Western world today, saying, "It is, roughly, an energizer which wakes us up and enables us to be a little more active, to go out and do things." More or less the same is true, he said, of nicotine, and of tea, which contains large amounts of caffeine. These stand in sharp distinction to many of the drugs commonly used for centuries in the rest of the world: cocaine, the opiates, marijuana, and heroin, which tend to make people drowsy and dreamy. "It would be interesting to know from a psychopharmacological point of view," he speculated, "what the daily use of these drugs by hundreds of millions of people might have done to Western expansionism and feelings of ability to shape the world."

Platt continued to his second point, which made a distinction between technology as a utility and technology as a determiner. Technology as a utility is something that is useful to the individual; hence it is liked. Technology as a determiner is something imposed on the individual by somebody else. Platt proposed telephones and cars as illustrations of his point. When you have a telephone and can use it to call your friends, you like it; it is technology as a utility. In contrast, when a telephone call from a prankster awakes you in the middle of the night, you resent the telephone; it is being imposed on you. Similarly, when you first get a car, you can use it to go places; you like it. However, when you are on the freeway in the middle of a traffic jam, you resent it; it has become a determiner of your time.

Platt extended the line of argument to science and technology in general. The switch from utility to determiner, he said, is essentially a case of numbers. In small numbers, "the cows on the common" are a utility, but in large numbers, the cows destroy the common and there is a net loss. Platt averred that the problem today is not so much that the nature of science, or of technology, or even of their respective individual returns, has changed, but rather that mass consumption has accelerated their use to the point at which such limited resources as time and space have been threatened. Our twentieth century problem is not due to a change in the characters of science or technology, he concluded. It is due to a correct perception that multiplication of some of these utilities makes them into determiners, which are then seen as hostile.

David Landes replied that he agreed wholeheartedly with Platt on his observations concerning psychopharmacology. In support, he noted that "the principle commodities in Europe's import trade from 1500 to 1800 were precisely what you [Platt] spoke of: tobacco, coffee, tea, and sugar." He elaborated on the nature of sugar, which energizes people though it is technically not a drug. It was, by Landes's description, "the big fix." These commodities, Landes continued, began as precious luxuries and became, by the 1800s, everyday commodities for all Europeans, even poor ones. Landes mentioned, however, that there has been very little research into the effect of these products on people who take them in "nonpathological" doses. Landes's own conclusion was "that it made urban and industrial life—and its pace—possible in a way that otherwise would never have been."

In response to Platt's proposition of technology as a utility versus technology as a determiner, Landes affirmed, "That is the whole story. We love it as long as it suits our needs and circumstances, and we get very upset by it if it deprives us of a job or seems to run things." Landes cited the ideas of David Dickson, an Englishman who theorized that the telephone is a product of an individualistic society; it discourages group interaction and promotes interaction between individuals. Landes concluded that sensitivity to technology as a utility versus technology as a determiner varies from individual to individual and from society to society.

Karl Deutsch continued with the theme of psychopharmacology and suggested that perhaps drugs function in various societies as mood amplifiers, accentuating existing social mores and cultural traits. To some extent, he said, each society, East and West, uses a little chemistry to make itself more like it was to start with.

Landes agreed, saying that the demand for certain drugs by certain cultures was not an accident. "It is not like the drug came in and made people this way," he observed. Landes pointed out how interesting it is that Western society, which has a tradition of using stimulants and alcohol—the latter of which, though technically a depressant, is used more by our society as a stimulant—is now turning to "Eastern" drugs. He wondered what such a transition signaled about the change in the character of our society.

Robert Lopez said that this issue showed that historians are not always in agreement, explaining that he "would not buy the idea of geographic circumstances." All historians of all civilizations, he maintained, have found reasons why the climate and geography of their respective countries are the best and, hence, that theirs is the "master race."

Lopez turned to the topic of the technological variations between Europe and China. He believed that Europe did not learn as much from China as it could have; China was just as highly developed as Europe until the Mongolian invasions, which brought catastrophe upon China. What he did believe to be the substantial difference between China and Europe—and the reason for Europe's eventual domination—was fragmentation. Many possibilities for individualism and variety, which were stymied elsewhere, could occur in Europe because of its fragmentation. On this point he agreed with Landes. Lopez did not, however, agree that Protestantism was also a cause of European development. The reasons why he believed Europe to be unique—but not necessarily better —were the tremendous importance of its city-states on the one hand, and its modification from antiquity to the Middle Ages on the other.

Lopez also mentioned the medieval tradition of time: it was vital to Europeans that they did not waste time, a trait emanating from certain structural prerequisites that did not exist elsewhere. In contrast to conditions in Europe, Lopez detailed, it was impossible in China to develop a progressive science as a result of the total control by the central authority. "We need anarchy, individualism, and struggle for diversified social progress," Lopez maintained. He concluded by saying that, if history teaches us anything, it succeeds in doing so with the help of small statistical samples such as case studies. He cited classical Athens, medieval Florence, and modern Paris as examples of city-states that were large enough for variety and yet small enough to allow for contact among individuals.

Departing from the strictly historical framework of the discussion thus far, Thomas Trautner made three observations. First, he suggested that one of the causes for disillusionment with science "did not develop from

science itself, but rather from the contemporary predicament of science."
He illustrated his point with an example from medicine. At the turn of
the century, there were tremendous advances in bacteriology which led
to a reduction of mortality and to longer life; scientists and doctors were
able to wipe out infectious disease to a great extent. However, this in turn
created new problems: for example, the rise of geriatric diseases. Thus
optimism, justified at the onset of the fight against infectious diseases,
evolved into pessimism, not because the discovery of microorganisms was
bad, but rather because it generated secondary problems. Second,
Trautner spoke to the interrelationship between political systems and
science. He mentioned the genetic crimes of the Nazis and the fact that
they were based on misconceptions of the uses and nature of science.
Trautner believed that these two phenomena contributed to general
disillusionment with science.

Trautner's third point related to the fear of science; such fear, he said,
was manipulated fear. Today's fear of genetic recombination, he maintained, was a result of media manipulation, since "this area of research
has no danger beyond that of handling pathogenic microorganisms." He
elaborated on this point by comparing the fear of genetic recombination
with the general acceptance of antibiotics, which he felt *are* dangerous.
Following the initial euphoria and consequent ubiquitous use of antibiotics, their widespread prescription led to abuses which have worked
against nature and have artificially created drug-resistant bacteria. This
is a highly dangerous development because, in many cases, real or
probable, we cannot treat certain illnesses with normal antibiotics. He
cited the cases of severe epidemics in Mexico and San Salvador, which
had such disastrous effects because bacteria had developed resistance to
antibiotics.

Landes questioned Trautner about the parallel between genetic engineering and antibiotics. Trautner corrected Landes's perception of what
he [Trautner] had said by reiterating, "There is no indication that the
kinds of things that are being done in genetics are dangerous." Landes
wondered why Trautner did not apply the same argument to antibiotics.
There followed a general discussion concerning Trautner's views on
antibiotics, in which there was much contention over his assertions.
Another scientist, Hans Wolfgang Levi, said that we must ask ourselves
the question: is it distrust in science or a distrust in scientists that is in
play? He said that people would rather trust what the media say than
what the scientist says, and asked why this was the case.

Joseph Weizenbaum commented, in this context, that there was
another possibility. "Science may be all right," he said, "and scientists
may be all right, but one should not introduce such powerful machines,
mechanisms, or devices into this society, which is essentially like a

fourteen-year-old child: reckless. For example, a Ferrari is a perfect, wonderful machine in the hands of a responsible adult. It is deadly in the hands of a fourteen-year-old child. Are we against cars, car manufacturers, Ferraris, or drivers? The answer is that we are against none of these." It depends on the configurations and their interaction. Analogously, there is nothing wrong with science or scientists, but rather with our contemporary social conditions under which science operates: in terms of self-control, society can be likened to a fourteen-year-old child.

Continuing with the topic of the fear of science, Peter Weingart raised the example of the catastrophe at Three Mile Island. He noted that on the day of the disaster, the Associated Press changed its headline twenty-eight times. This went on throughout the first week of coverage and reached its conclusion with *Newsweek's* statement that "the first casualty of the Three Mile Island accident was scientific credibility." He felt that Trautner's contention that the media manipulated public attitudes was too simplistic. There is manipulation involved, but much of it is initiated within the scientific community itself.

Levi replied that the loss of scientific credibility after the Three Mile Island incident was due to misunderstanding by the public. Nuclear scientists never said that there was *no* danger—which is what the public understood them to say—but rather that the probability of danger was very small.

Bruno Fritsch directed two questions at Trautner. He asked first if a cost-benefit analysis today would show that antibiotics should not be used, and, second, whether scientists foresaw the problem. Trautner replied that scientists did not foresee the danger of antibiotics, but it became clear to them later. Many attempted to seek controls, but by then the use of antibiotics was tied to habits and economic interests. The situation today is slightly improved; there are laws regulating the use of antibiotics in many countries. In reply to the second question, he said that a cost-benefit analysis would be enormously difficult.

I. Bernard Cohen pointed out that in the eighteenth century there was a consensus in the scientific community, partially because science did not concern itself with large-scale effects or large-scale policies. However, from World War II to the present time, in which science has assumed such a role, there have been eminent scientists available to defend almost any point of view. "I do not know of any issue on which there is absolute unanimity in the whole scientific community," he noted. In light of the diversity of opinions within this community, he said, the public is put in a difficult situation. On the question of manipulation, Cohen too felt that Trautner's idea was overly simplistic; there exists a great pluralism within the scientific community which defies simple categorization.

The point was also generally noted that, historically, public opinion about scientific accomplishments has traditionally been ambivalent. Technological developments have always been greeted with expressions of enthusiasm together with voices of alarm. Furthermore, some skepticism was voiced as to the role of drugs in the development of European innovation and technology. More relevant explanations seemed to be the European interest in reading and the invention of the printing press.

Landes commented that although printing was an important contribution, it came after Europe had already shown many signs of innovation and technical progress. Though the subject of why Europe became the leader in scientific innovation was not the topic of the conference, Landes held that it was crucial to place the issue of fear of science versus trust in science in the larger context of the European tradition of commitment to scientific development, which is at least a thousand years old. Gerhard Leminsky suggested that it might be useful to add a structural component to Landes's presentation; social structure, he commented, has been a decisive influence on the direction and the content of science.

Commenting on the instrumental aspects of science and technology, Wilhelm Kewenig said, "If one talks about fear of science in our time, it is not so much fear of science, but fear of the misuse of science by political or economic power." Kewenig characterized this way of posing the question as a more accurate application of the theme of the conference to contemporary conditions.

A study performed by Helga Nowotny was discussed by Elisabeth Helander. Nowotny examined the role of experts in the public discussion preceding the plebiscite concerning the use of Austria's first nuclear power plant, which was constructed near Vienna, in Zwentendorf. The Nowotny study showed that, when the experts were pressed during public debate, they made statements that went beyond their scientific knowledge. They seemed to feel compelled to depart from their disinterested presentation of objective scientific knowledge to a position of advocacy. Another point made by the study was that the strongest proponents of nuclear power were highly specialized nuclear engineers; those who were opponents tended to be scientists with wider interests and experiences.

Jean Stoetzel noted that the historians had overlooked the role of psychology. In pointing out the rise of opposition to psychological tests, surveys, and polls, Stoetzel drew a parallel between the distrust in science and the growing distrust in psychology.

By suggesting that mankind "struggles to cope," Deutsch approached the issue from a different perspective. He proposed that we perhaps should ask if science has helped man in his attempt to "cope" better.

Landes concluded by briefly addressing Stoetzel's comment. Fear of the human, or behavioral, sciences has certainly added to the mistrust of sciences for the reasons which Stoetzel mentioned: for example, scandalous results of polls and surveys. Because of the political implications of such practices, some people are not prepared to accept their results at all, even when they are being told facts and not conjectures. Landes illustrated this total skepticism on the part of some people vis à vis polls in particular and science in general, with an anecdote. In the course of a discussion with the wife of a colleague, Landes brought up the controversial issue of the validity of standardized tests; there are significant differences in performance by ethnic origin and by sex, he told her, which lead many to believe that the tests ought to be abolished. The woman maintained that there is no way to make an unbiased test. Landes suggested that certainly the following question was unbiased. A picture of a dog with only three legs is shown. The person taking the test is asked to identify what is missing. Landes told her that it would seem to him that it did not matter whether the person was a man or a woman, black or white, upper or lower class: the person should be able to answer correctly.

The woman, however, said that she knew children who had lived in neighborhoods where they may well have seen three-legged dogs and may have thought it was normal. "Now I do not know what you do," Landes exclaimed to the conference participants, "with an answer like that!" Such an attitude, he said, shows a refusal to accept the implications of social and behavioral sciences. "It is one thing to tell people that nature is a certain way," he observed, closing the discussion, "but it is another thing to tell them that human beings are a certain way!"

PART II

The Present Situation: Big Science and the Big Fear

Chapter 9

Some Notes on the Fear and Distrust of Science

Harvey Brooks

DOES IT EXIST? WHAT IS IT?

The evidence on public attitudes towards science and technology is very mixed. Furthermore, one cannot talk about a single public; there are many different publics—the "man in the street" with a low level of consciousness about science, the literary intellectual of the mass media, the average educated man largely represented in professional groups, the disadvantaged individual from a minority group, today's woman, the young person. Each of these has distinctive attitudes.

There is no question that the automatic beneficence of science, and of technological advance, is much less taken for granted than has been the case throughout most of American history. The rise of the environmental movement, and of technology assessment, reflects a growing consensus that society must be much more selective and discriminating about the technologies that it chooses to apply on a large scale. Although this critical attitude focuses more on technology than on science, it has begun to hit science in those areas where the distinction between science and technology is least easy to maintain. Examples are in recombinant DNA research, where the laboratory techniques used in research can be readily visualized as translating fairly directly into industrial processes, and in synthetic organic chemistry, where again the relationship of laboratory methods to large-scale applications is very evident.

Another area that has aroused widespread fear is nuclear fission. However, in this domain it appears that the public has been more ready to accept the separation of pure research and technology, since nuclear physics research has not been subject to criticism to the same degree as molecular genetics. Although both nuclear weapons and nuclear power derived originally from fundamental research in nuclear physics, the connection between the present frontiers of research in nuclear and particle physics and the engineering applications of nuclear fission and artificial radioactivity is much more remote than in the case of molecular biology and genetic engineering.

Despite the fact that the scientific community perceives a public turning against science, little evidence for this can be found in public opinion surveys. Scientists as a profession stand very near the top of the scale in public respect and status. Scientists are for the most part treated with greater deference and respect as Congressional witnesses than almost any other professional group. Politicians and administrators increasingly look to scientists and the organized institutions of science to provide advice and data as input to public decisionmaking on issues such as nuclear safety, arms control, environmental protection, energy policy, toxic chemicals, and a variety of other public issues of high technical content. This is more true today than it was in the flood tide of scientific prestige which followed the public scare over Sputnik early in the 1960s. Thus it is really hard to say whether what is observed is merely part of the generally rising distrust of institutions and authority which has been pervasive in all the Western democracies since the late 1960s.

INTELLECTUAL HERITAGE OF ANTISCIENCE

In the history of the last three centuries, there is a long heritage of distrust of science and technology among humanist intellectuals. This is closely linked to disenchantment with the side effects of modernization and the industrial revolution, and what is seen as the increasing alienation of man from nature and from his own fellows. Thus feelings against the scientific tradition are nothing new. What *is* new is that much higher literacy and levels of education, plus the new communications media, have facilitated a far wider diffusion of antiscience and antitechnology—enough to make them a politically effective force in many of the industrial democracies of the West, as evidenced by the various "green" parties in Europe and the hundreds of environmental action groups that have become an effective political force in the United States.

While it would be unfair to identify all of these groups with anti-technology, much less with distrust and fear of science, there has been a strong undercurrent of antiscientific ideology, especially among the leading figures. At the same time, many of these groups invoke scientific data and analysis to back up their opposition to specific development projects such as nuclear power plants, large dams, the Alaska pipeline, supersonic aircraft, microwave ovens, and recombinant DNA research—to name just a few of the targets of environmental attack. Simultaneously they also invoke science in support of newer and supposedly more benign "appropriate technologies," usually associated with small scale and the use of renewable rather than nonrenewable resources. To the extent that this is so, these people are trying to distinguish between science itself and its contemporary applications.

Nevertheless, many of the most articulate environmental spokesmen do express a new vision of society in which technology as traditionally conceived plays a declining role. Many also focus their attack on those technologies of recent vintage which are most obviously derived from science rather than from more traditional art and craftsmanship—computers, nuclear energy, sophisticated military weapons, intensive mechanized agriculture, synthetic chemicals, high technology medicine and pharmaceuticals, weather modification, microwaves (but not television), supersonic aircraft. In the minds of these critics, science has become subservient to these technologies and to some extent has fueled their application and diffusion; therefore science can no longer claim its distinctiveness from technology or assert its neutrality with respect to the political controversies of our time.

Indeed the arguments used by basic scientists in the 1960s to defend and justify generous public support of fundamental research may now be boomeranging. These scientists attributed to pure science as a virtue the very derived technologies which the critics now fear and reject. Thus science itself is feared because of the increasing power for either good or ill which the technologies derived from it can create. Science, especially "pure science," which was seen as entirely benign a century ago, is seen today, even by many scientists, as ambivalent in its effects on society, increasingly prone to misuse by governments and large corporations (never small corporations), and increasingly liable to generate unexpected side effects which are harmful. Thus the scientific community finds itself on the horns of a dilemma. The more it defends the support of pure science because of its ultimate utility to society, the more it opens science itself to the same criticism as technology and hence to a public backlash which may defeat the very purpose for which science was defended.

ROLE OF SCIENTIFIC COMMUNITY IN DISTRUST OF SCIENCE

It must be admitted that much public fear and distrust of science has been fostered from within the scientific community itself, not by the lay "enemies" of science. If one looks at the sources of opposition to large technological projects, one almost always finds that the original impetus came from scientists, usually a minority who chose to appeal to the public over the opposition, or, more frequently, the indifference, of the majority of their colleagues. Whether these dissident scientists were right in particular instances is less important than what their cumulative impact on public attitudes toward science was. The resulting dissension within the scientific community led the public to see that community increasingly as another "establishment," more analogous to business, labor, and farmers, than as a disinterested "apolitical elite" devoted to the larger public interest. I happen to feel that most of the concerns expressed by dissenting scientists ranged from greatly exaggerated to downright trivial or wrong-headed, but even in their exaggeration they frequently called attention to considerations which the establishment had overlooked or had not taken sufficiently seriously. Thus, the critics generated a dialectic which has often in the end proved constructive, though it sometimes delayed or stopped projects which were probably on balance in the public interest.

Physicists, for example, spearheaded the initial opposition to nuclear power. Initially it was their intent to advocate a "go slow" policy until certain faults they saw in the technology could be properly studied and corrected before further deployments. Later, as their criticisms caught on with the public, many of them became captive to the dynamics of the antinuclear ideology and went much further in their opposition, advocating a moratorium on new plants and the shutting down of existing ones.

It was molecular biologists who first raised the alarm over research with recombinant DNA, ecologists who provided much of the scientific underpinning of the environmental movement, physicists who largely organized the defeat in Congress of the American commercial supersonic transport program, computer experts who warned of the problems of computers and privacy, chemists who sounded the alarm about toxic chemicals as an occupational and environmental hazard, safety engineers who provided the basis for Nader's crusade on auto safety, and biologists who aroused fears (totally unjustified, in my view) about the possible adverse physiological effects of low-level microwave radiation as well as other nonionizing radiations. It is extremely doubtful whether any of these subjects would have become significant public issues without the stimulus of scientists, and without at least the appearance of some supporting scientific evidence, however dubious its validity.

Dissension within the scientific community about the dangers of specific technologies and areas of research has had a dual effect on public attitudes. It has certainly confused the public on such issues as nuclear power, offshore oil exploration, the time scale on which solar energy can help our energy problems, or the potential health hazards to the public of research using recombinant DNA techniques. Because of this, it has eroded public confidence in the infallibility and the objectivity of science. Much of this confidence was misplaced in the first place, because the public found it difficult to distinguish between proven scientific fact and hypotheses based on partial evidence and intuitive scientific judgment. Furthermore, many controversies involving technical considerations turn on predictions about the future or predicting the future social and environmental consequences of specific policy choices. These judgments about the future are inherently less factual and frequently involve knowledge outside the domain of expertise of people who understand the technology itself. However, the controversies may also increase confidence in scientists, through helping the public to perceive them as disinterested defenders of the public interest regardless of the consequences to their own interests. Surveys of public opinion about nuclear power, for example, show that scientists still have the highest credibility with the public on this issue, even when they disagree; the public still expects them to reach a consensus and, when they do, is ready to follow them.

On major public issues, advocates on both sides appeal to scientific authority in support of their positions. Congressional committees seek out scientific witnesses either to buttress predetermined positions or to establish a data base for policy. A decade ago Congressman Emilio Q. Daddario urged scientists to enter more actively into the public rough-and-tumble, to come out from behind their screen of apolitical objectivity and speak out as advocates of public policy rather than as mere resources for the politicians. Daddario's wishes have been largely realized in the last decade, but it is difficult to decide on balance whether this development has increased or decreased public confidence in science.

PUBLIC SUPPORT OF SCIENCE

With respect to government support of basic and academic science, the past decade has seen an absolute decline in real terms, although there has been some recovery in the last few years. In fact the decline has been largely the result of inflation, not of a decrease in the allocation of tax dollars. Nevertheless, many in the scientific community see the declines as evidence of public disenchantment with science, even of an antiscience trend. However, I think it is more likely interpretable as the result of the competitive pressures of new budgetary priorities, especially social welfare and income transfer programs, whose relationship to science is much weaker than earlier government priorities in defense, space, or curative medicine. Indeed, the recent reversal of the declining trend can be seen as partly the result of the emergence of energy and the environment as major national priorities, and the fact that these have much higher technical content than do income-transfer programs.

It is true that politicians' skepticism of the value of nondirected research has always been endemic in Congress, and emerges in periodic actions, such as the famous Mansfield amendment mandating the direct military relevance of military-supported research, or the recent attacks on the integrity of the peer review process for selecting research projects for support by the National Science Foundation. Nevertheless, there is little evidence that this skepticism, and the public distrust which may underlie it, is any greater than in the past.

It is a fact that, compared with the 1960s, the scientific community is increasingly viewed by politicians as another interest group, clamoring in competition with other groups—the farm lobby or the maritime industry, for example—for federal subsidies because of its asserted vital contribution to the public interest. Certain spokesmen of science have in the past made promises regarding the public benefits to be realized from the support of particular fields—benefits that have been slower in coming than was implicitly if not explicitly understood in the original advocacy. This has certainly fed the current skepticism, particularly in the medical field. The fact remains, however, that Congress still continues to appropriate more funds for biomedical research than have been requested by the Administration (regardless of political complexion), as it has done for the past twenty years. In the Fiscal Year 1980, the budget of the National Science Foundation (NSF) passed the billion dollar mark for the first time, and a considerably larger fraction was for basic research project grants than in the supposedly more prosperous times of the mid-1960s. Over the long pull, the basic research part of the NSF has fared better politically than supposedly more appealing programs such as science education, research applied to national needs (RANN), weather modification, science development programs, or institutional formula grants.

The question that is being discussed in the scientific community today is whether the national consensus which apparently crystallized, right after World War II, behind the famous Bush report, *Science, the Endless Frontier*, is beginning to disintegrate. There is also a question of whether there ever was such a consensus or whether science has merely ridden on the coattails of successive national priorities, from the arms race with the Soviet Union, through the war on cancer, to the energy crisis. Today there appears to be wider agreement on the importance of a sound basic research program for the future health of the American economy than at any time in the past fifteen years; in a time when the role of the federal government in the national economy is coming under increasing criticism, however, it is hard to foresee future trends. The support of science seems likely to be affected more by general fiscal policy and by general public attitudes towards growth of the public sector than by any skepticism or revulsion directed specifically at science.

SCIENCE AND RELIGION

Of some interest has been the reemergence of organized religion as a spokesman for antiscience. This is particularly true of recent pronouncements of the National Council of Churches and the World Council of Churches. While many of their statements have focused on the nuclear power issue, they seem to identify nuclear power as a symbol of the reductionist, rationalist, and manipulative style of thought which they attribute to modern science. Science's attempt to exclude wishful thinking is misinterpreted as a deliberate effort to exclude all human values from life, particularly as scientific methodology asserts its claim to contribute to the rational understanding of wider and wider domains of human experience.

The popularity of various forms of oriental or Christian evangelical religions among youth may be another symptom of a revulsion against science. Still, it is interesting to note that virtually all these religions, which claim to embrace all of human experience in a holistic sense, also attempt to invoke scientific evidence and contemporary scientific concepts in support of their claims. Indeed a recent news report in *Time* magazine speaks of a rapprochement between a small group of theoretical physicists and certain Indian religious ideas. One could say that this represents the sincerest form of flattery of science, and is thus evidence that the prestige of science requires even its enemies to disguise their critiques of modern science in the language and the concepts of science. Such a technique, however, has often been characteristic of religious controversy in the past, with each new religion claiming to be a purified form of an established religion that has become corrupted by worldly success. In other words, invoking the language of science is not a tribute so much to the public respect in which it is held as to its worldly success as a part of the modern establishment.

IS FEAR OF SCIENCE JUSTIFIED?

One cannot discuss fear and distrust of science without facing the question of the degree to which this distrust may have been earned either by the social consequences or the behavior of the scientific community as a collectivity, or of scientists individually. In a world in which egalitarian ideology flourishes and is fashionable among intellectuals, science is frankly elitist, with stringent criteria for membership in the circle of the elect. The scientific community defends its elitism on the basis that it is a meritocracy, based on universalistic criteria where merit is assessed independently of any personal characteristics or affiliations. Research on the sociology of science for the most part supports this claim, especially as far as the reward system within science is concerned. Nevertheless, the scientific community is embarrassed by the underrepresentation of certain groups within its ranks—women, blacks, and Catholics, for example. Whether this is due to discrimination, or, more specifically, to the perhaps unconscious application of nonscientific criteria to the reward system of science, is difficult to prove one way or the other. It is hard to make a case that science is worse in this respect than other high status professions. It is probably a good deal better, but it also makes larger claims to universalistic standards. Usually, apparent discrimination is attributable to aspects of socialization that occur prior to entry into the scientific community, but it is difficult to prove that this is the whole story, especially in relation to the underrepresentation of women.

Closely related to the problem of elitism are the closed decisionmaking processes of the scientific community. Scientists insist that only scientific peers are competent to make decisions about the substantive direction and priorities of research within a field. They may somewhat grudgingly accept the legitimacy of lay judgments about broad priorities within science as a whole, admitting that scientists themselves have been unwilling to pass judgment in public about the work of colleagues in fields very distant from their own. *De facto* decisions about the relative investment in, for instance, biomedical research compared with nuclear physics have been made politically, with spokesmen for competing sciences lobbying the political process.

Scientists often view themselves as a persecuted sect whenever confronted by lay criticism or threatened with regulation, especially when it entails the delegation of significant decisionmaking power to nonscientists. Scientific organizations expect to be managed only by individuals who have first made their mark as contributing research scientists in their own right. There have been exceptions to this, such as the general acceptance of Dr. James Killian as the first science advisor to President

Eisenhower, but this is rare. A case can be made that research scientists do indeed require a good deal more internal autonomy than do most other professions—first, because they do not interface directly with the public (unlike doctors or lawyers), and, second, because the necessities of specialization place understanding of the day-to-day strategy of research beyond the capability of laymen without a great deal more time and effort than they are willing or able to afford. Most working scientists are contemptuous and suspicious of the mass media and extremely critical of colleagues who "go public" before their ideas have been subjected to collegial evaluation through publication in a professional journal. All of these opinions and practices can be defended as ways of protecting the autonomy of the selective processes within science—of defending the integrity of the scientific process—but in an age where public support is so essential, and in which there is little market test of the product outside the internal "market of ideas" within the scientific community, uncommunicativeness with the media, however justified, breeds public distrust and suspicion.

In all probability, nothing has engendered fear of science as much as some of its major by-products: weapons of mass destruction, toxic and polluting synthetic chemicals, manmade radioactivity, depersonalizing applications of computers, and large-scale, impersonal organizations made possible by modern information technology and communications. Even the beneficently intended applications of science, such as the green revolution and civilian nuclear power, have come in for attack because of their potential side effects sometime in the future.

These concerns raise the question of the responsibility of science or scientists for the applications made by society of the results of their research and of their special knowledge. Even admitting such responsibility, how can and should it be exercised in practice? Should scientists collectively withhold their services from activities and organizations whose purposes or activities they consider to be adverse to the welfare of humanity? If scientists in some collective sense decide not to carry on certain activities or not to serve certain organizations, does a majority, or even a dedicated minority, have a right to enforce its collective view on all other members of the scientific community? For such a purpose, who belongs to the scientific community? To what extent do scientists as a group have a right or a responsibility to enforce their views of what is in the public interest against the preferences of society as expressed through normal political institutions and processes? Do they discharge their responsibility by working through normal political channels, or merely by expressing their views in public? If participation or nonparticipation, speaking out or keeping silent, is left to individual conscience, will not the behavior of the least responsible and sensitive members of the community determine what happens? Science is not a profession like law or medicine, where it is possible to exercise discipline over a well-defined membership. Science is, in principle, an open community in which anybody can become a member merely by virtue of general recognition of his contributions to knowledge. He does not have to be licensed or approved through a formal process to practice his profession, although his activities may be constrained by the degree of recognition he is actually accorded by his peers. However, whatever the responsibility of scientists, society tends to hold them responsible after the fact for the uses made of their science. The fact that few scientists may have foreseen the problem is quickly forgotten. Thus attitudes towards science are inevitably contaminated by attitudes towards the results of scientific applications, however remote the connection may be.

Indeed the remoteness of the connection between science and its ultimate applications is a major part of the problem of assessing scientific responsibility. Is the discovery of $E = mc^2$ to be held responsible for the atomic bomb, or is this the case only if the discoverer saw the potential application of this relationship at the time? Even if the discoverer did foresee the possible application, how was he to weigh the potential benefits against the potential risks at some remote time in the future? Not only are these risks and benefits largely unknown and unknowable, but also they depend strongly on the state of society at the time applications first become possible. The application of science is determined largely by social processes external to science, although scientists may play a more or less influential part in these processes. It is true, for example, that prominent scientists, recognized leaders and opinion-formers among their peers, played a key role not only in the development of fission and thermonuclear weapons but also in their advocacy in the political arena. Teller, Oppenheimer, Lawrence, and others hardly constituted an "apolitical elite" who could divorce themselves from the applications of their discoveries made by society. Yet their advocacy of nuclear weapons was sincere, and entirely in what they believed to be the public interest, heavily conditioned by the general political climate of the time. Even with today's wisdom of hindsight, it is possible to debate whether the existence of nuclear weapons has prevented major war for more than thirty years, or whether it has only postponed conflict at the price of making it infinitely more destructive, perhaps fatal to human survival on the planet, when it finally comes. Only time may disclose the answer, and even then we will never be certain.

Many of the scientific and technological developments that arouse most public apprehension—nuclear weapons, nuclear power, the green revolution—can be regarded as buying time for humanity to make more far-reaching changes to assure its future. The world dilemma at present is in no small part due to the application of public health measures in a poverty-stricken world in which fertility was uncontrolled. Yet virtually nobody would argue that, once it came within our power to save lives on a large scale, we would deliberately withhold this power. In this sense science has saved far more lives than it has destroyed by instruments of mass destruction, and yet, ironically, this humanitarian accomplishment may prove more dangerous to the human future than the destructive achievements. Nuclear power may buy time to develop more benign, indefinitely sustainable energy sources to replace exhaustible fossil fuels. The green revolution may buy a generation's span to get the population explosion under control without the assistance of mass starvation or thermonuclear holocaust. However, each of these developments might also make things worse if humanity fails to use the time wisely. Nuclear weapons make destruction more terrible if deterrence fails. Nuclear power may increase the danger of war through the proliferation of nuclear weapons among hundreds of assertive sovereign nations. The green revolution may merely allow population to continue to grow, making the ultimate famine far more disastrous and sowing the seeds of world political disintegration through enhancing the inequalities in developing societies and the rich-poor gap in the world as a whole.

The point of the preceding discussion is to illustrate that it is not realistic to expect organized science or scientists to foresee all the complex interactions between science, technology, and society which determine how scientific discoveries will actually be applied to influence social, political, and economic developments. Indeed, foreseeing the consequences of technical developments is not often within the competence of the authors of those developments. It usually requires quite different knowledge and skills which, however, have to be somehow synthesized with understanding of the technologies themselves. Thus it seems hardly fair to hold scientists responsible or accountable for consequences that they may be ill-equipped to foresee.

While it may not be reasonable to hold scientists strictly accountable for the ultimate consequences of their work, it is, however, essential to acknowledge the political role of some scientists, whether one thinks that their main influence was for good or for ill. Certainly the role of prominent scientists in "selling" various technologies has contributed to the public fear of science and the identification of science itself with the technologies involved. Conversely, if environmentalist opposition to nuclear power and to other future energy supplies results in severe economic and personal hardship, while solar energy and renewable resources do not prove the panacea their advocates believe, there may be a severe future backlash against science because of the part played by prominent scientists in providing scientific legitimacy to these movements. The participation of scientists in political advocacy certainly invites future distrust of science if they turn out to be wrong, and all of science will to some extent share the resulting opprobrium.

It is difficult to imagine, however, a scenario for science or scientists which would have a different outcome in terms of public confidence. Indeed in the political debates surrounding public decisions about various technologies, scientists have been aligned on both sides, and in a division not too different from that of the generally informed nonscientific public. It is difficult to see how it could be otherwise, because in fact many of the issues hang more on value judgments outside of science than on the scientific facts themselves. Much of the difference, for example, lies in the essentially moral judgment as to where the burden of proof should lie with respect to the safety of various activities.

SCIENTISTS AS AN ESTABLISHMENT

Another factor affecting public attitudes towards science is the organizational affiliations of scientists and engineers. About 70 percent of all scientists and engineers work for private industry, many of them under contract with the federal government. Not more than 12 percent work for universities and independent nonprofit institutions, and even this segment has its research predominantly funded by the federal government. In addition, many academic specialists work as consultants to industries to which their expertise is relevant. Scientists and engineers are thus salaried employees of a gigantic $54 billion a year R&D industry supported about half by corporations (mainly large ones), and about half by a handful of agencies of the federal government, of which the Defense Department and NASA account for well over 50 percent. During a time of increased public distrust of institutions and authority generally, the surprising thing is not that scientists are distrusted, but that there is still such a comparatively high level of trust, as indicated by some of the examples at the beginning of this essay.

To a large extent, I think that the distrust of science that exists is not a distrust of science itself, or of the scientific method of knowing reality, but a distrust of the institutions which are the main employers of science and scientists. The fear is that most scientists may be the intellectual captives of the interests and purposes of these institutions rather than the disinterested servants of the public interest. In fact, this is probably one of the reasons why "public interest scientists" who speak as dissenters and critics of the establishment receive a more sympathetic hearing from the public than the scientific merits of their arguments might otherwise deserve.

One of the responses to this fear of the captivity of scientists to special interests is the move to provide public financing for intervenors in hearing procedures involving the licensing or permitting of various technological projects. The purpose is to permit them to finance independent scientific analyses of the projects proposed, in order to provide a counterweight to the expertise to which the promoters of the projects have access. This procedure has been more extensively adopted in Europe than in the United States, especially in the Netherlands and the Scandinavian countries. In principle, it should increase public confidence in science and technology, but it may also generate confusion and indecision. In the American system particularly, the ability of intervenors to create costly delays to private investors may largely offset their lesser access to expertise, so that public financing would tip the scales too far in favor of inaction. On the other hand, if public financing of intervenors is used to develop a more knowledgeable public before actual confrontation occurs, it may have a constructive effect in facilitating rational compromises before the various parties have become entrenched in fixed positions.

THE OTHER SIDE OF THE COIN: MISPLACED FAITH IN SCIENCE

There is an ambivalence about public attitudes toward science. Although there is a rising level of distrust for the reasons I have outlined above, there seems also to be an increasing belief that scientists can accomplish almost anything if the public is willing to hold their feet to the fire. This faith has in part been created by some of the spectacular accomplishments of science and technology during and after World War II. The Manhattan Project and the Apollo Project are most frequently cited as precedents for what could be accomplished if only technical resources could be mobilized on a similar scale for the solution of some of our current social problems, particularly those of energy. This has been referred to by Richard Nelson as the "moon-ghetto syndrome," and has led to the mounting of successive technological campaigns for the solution of this or that problem, ranging from poverty and cancer to solar energy and aging. The legislation establishing automobile emission standards in 1970 was based on the premise that engineers could come up with an immediate alternative to the internal combustion engine if the automobile companies were faced with sufficiently stringent penalties for noncompliance with tough standards on tight deadlines. Accomplishments since 1970, while falling far short of the original standards, considerably exceeded the expectations of most experts.

Today, one finds similar arguments being used with respect to energy conservation and solar energy. What is not appreciated is that the atomic bomb and the moon landing were both largely achievements in "pure technology." They met well-defined goals in a political context where cost was a negligible consideration, and where the final product did not involve the large-scale deployment or reproduction of the original technological product. The Manhattan Project resulted in the dropping of two bombs, and the Apollo Project culminated in a dozen or so moon landings. There was no product that had to be mass-produced on short notice and then operated reliably for many years by people with much lower levels of skill and knowledge than the scientists and engineers who developed the prototypes. The first atomic bombs cost about $2 billion to develop while the entire Apollo Project cost about $24 billion. However, to achieve either the energy supplies necessary after the year 2000 or even the efficiencies in energy end-use necessary to reduce the need for additional supplies, will require national investments that are measured in trillions rather than tens of billions of dollars. The technological achievement is only the first, and relatively minor, step in the solution of the overall problem.

Thus the paradox is that distrust of science and faith in science go hand in hand. The public feels it can stop particular technologies because, somehow or other, the technical community can always come up with alternatives, and technology can be channeled with money alone or with sufficient political will.

The other side of the coin is that the problems facing humanity in the next fifty years appear irresolvable without a heavy input from science and technology. However, science and technology can only be a small part of the total solution; they are necessary but far from sufficient conditions. Yet if public distrust too greatly inhibits their development and application, disaster may overtake the human race regardless of what social and political actions are taken. Whether disaster will take the form of thermonuclear suicide, famine and epidemics on an unprecedented scale, or simply general social deterioration, violence, and terrorism, is difficult to predict.

Chapter 10

Discussion of
"Some Notes on the Fear
and Distrust of Science"

Summarized by Andrei S. Markovits and Karl W. Deutsch

Joseph Weizenbaum began the discussion, expressing regret that Harvey Brooks was not present to defend his paper since he, Weizenbaum, had "quite serious difficulties" with it. He enumerated his concerns with direct quotations from Brooks's paper.

Weizenbaum first took issue with Brooks's implied assumptions about public attitudes toward science, in the passage that began with the statement that dissent in the scientific community tended to confuse the public. Brooks contended, "Because of this, it has eroded public confidence in the infallibility and objectivity of science. Much of this confidence was misplaced in the first place because the public found it difficult to distinguish between proven scientific fact and hypotheses based on partial evidence and intuitive scientific judgment." By faulting the public only for failing to distinguish between fact and hypothesis, Weizenbaum maintained, Brooks implied that science was infallible. "Science, in fact, is *not* infallible, and objectivity is a myth," he noted, and went on to observe that, while Brooks could probably say in defense that the implication was unintended, clear writing is a standard that should hold for all scientists who address themselves to the public, especially in consideration of the effects of the mass media on public attitudes.

Note: Harvey Brooks, the author of the paper, was unable to attend the conference. His paper was read by Thomas Kusack.

Weizenbaum next quoted Brooks on the role of scientific research within society in general. Brooks wrote, "Today there appears to be wider agreement on the importance of a sound basic research program for the future health of the American economy than at any time within the past fifteen years . . ." Weizenbaum stressed that he did not disagree with the statement, but he found it interesting because it came from "a highly placed member of the 'big science' establishment of the United States." That such an individual would make this kind of statement indicates that science, especially "big science" within the context of the United States, is not a "disinterested" practice in which scientists follow truth wherever it may lead, but that it is an activity instrumental to society and its economy.

Another point of disagreement for Weizenbaum was Brooks's statement that the recent pronouncements by American and world religious organizations "seem to identify nuclear power as a symbol of the reductionist, rationalist, and manipulative style of thought which they attribute to modern science. Science's attempt to exclude wishful thinking is misinterpreted as a deliberate effort to exclude all human values from life . . ." Weizenbaum countered that Brooks did not define the issue correctly: his [Weizenbaum's] criticism of much of modern science is not that it makes a deliberate effort to exclude human values. To the contrary, the effort is unconscious, as well it should be: science must recognize that it shows only *one* facet, not all facets, of the world, and that other facets, such as values, must be included in the full social discussion of science-related problems.

Weizenbaum's next criticism was directed at Brooks's defense of "elitism" in science on the grounds that science is a meritocracy, employing standards of "universalistic criteria" and not of "personal characteristics or affiliations." Giving examples to the contrary from his own experience, Weizenbaum questioned this defense and concluded that, though we may be closer to a meritocracy today than we were thirty years ago or before World War II, Brooks's statement must be assessed a good deal more critically.

Weizenbaum took particular exception to Brooks's mention of the "depersonalizing applications of computers." He wished to call attention to the assumptions implicit in such a description, since it was his [Weizenbaum's] view that depersonalization is not the important issue in discussing the dangers of computer technology; the important point, to him, is our loss of control over computers and thus over the functions to which we assign them. He further criticized Brooks for stating, ". . . society tends to hold [scientists] responsible after the fact for the uses made of their science. The fact that few scientists may have foreseen the problem is quickly forgotten." Weizenbaum deduced that Brooks's contention is that if that fact were remembered, the scientists would be excused; this attitude, he said, is reflected in Weizenbaum's own profession by the belief that computer programs are so complicated that the programmers cannot foresee their behavior and that therefore programmers should not be held responsible for the applications of their programs. This belief, Weizenbaum maintained, should be challenged.

Finally, Weizenbaum noted with interest that even such a technological optimist as Brooks made allusions in his paper to the possibility of the world coming to an end. The World Council of Churches had met in the summer of 1979 at MIT, and at that conclave Weizenbaum had witnessed a "heavy air of emergency, of the near end of the human race." Even scientific optimists, he said, as represented by Brooks, are coming to share this sense, namely, that there is much more at stake in science than whether or not the National Science Foundation will continue to support a particular research laboratory at MIT or Harvard.

John Platt contributed two points concerning public reaction to science today. First, a great part of the fear and misunderstanding of science can be traced to mass higher education, especially in the United States. He estimated that nearly 60 percent of the American population proceeds to college after high school and ventured that this 60 percent thus has the confidence that it has been educated and can therefore hold valid opinions on such subjects as science. Yet the colleges—faced with this "brood" antagonized to science by the "horrible" methods of the elementary schools, which drive even the student who loves science, or poetry, or mathematics, to despise it after one semester—have steadily decreased science requirements to the point where a great fraction of those holding college degrees have had no exposure to science at all. Still, this same body of degree holders, Platt argued, claims that it has standing to discuss science because it went to college and is thus "educated." It has, in fact, no understanding of the nuances of science necessary to meaningful discourse about complex matters. This applies to public and even legislative attitudes in the United States.

Platt's second point concerned what he called a "reversal" of social lag; this was defined as the time it takes for a society to adapt to advances in technology as the older, resistant generation dies off, and the younger, accepting generation takes its place. Television, Platt contended, with its immediate sharing of information, opinions, and ideas, has provided for a reversal of social lag. Everyone, young and old, learns together, not from each other but from "the box." They are exposed, together, to the same levels of information, entertainment, public affairs, and global concerns. Education is speeded up because it no longer passes from one generation to another through the inherent resistance between the generations. Television has brought such a high rate of awareness that now societies can say immediately, "We don't like oil spills," or "We don't like nuclear disasters." Platt concluded that this "no-saying" to scientists by society is one of the principal characteristics of our times. As already noted, the abandonment of the supersonic transport project by the United States marked the first time in Western technological history that politicians had said no to a multibillion dollar technological juggernaut already in progress, on largely human and environmental grounds.

Science is instead steered in a different direction, Platt said. Society wants mass transit, greater food supplies, nonpolluting automobiles, and solar power. Scientists respond by saying, "Yes, it is probably possible, but it will take about twenty-five years"; the social lag of previous generations has been transformed into the "technical lag" of today. In this way, our society becomes cybernetic in nature: its multiple groups, the highly diffused networks of protesters and interactors—in "the boondocks, not Washington"—try to decide on a course for society based on its present ideas about goals. It is a laborious process that may take a long time, but people feel now that they can steer technology, rather than that they are being steered by it. "I think this is a healthy trend for the future," Platt offered, "and I think that Harvey Brooks's paper misses completely this information-communication aspect of the present transformation of our society. When you make hundreds of millions aware, suddenly they want to choose."

Elisabeth Helander suggested that science itself must approach the question of whether some fears of science are really justified, whether the public fear of science is automatic, or whether, in some cases, there might be good reasons for it. This process, she said, would lead to genuine self-reflection within science. Scientists might question the structure of science, or ask whether we have the kind of science we really need for the future. Brooks mentions briefly, she noted, the bias that results from the link between science and the establishment, which gives rise to opposition to science by public interest groups. Is it enough of an evaluation of science when "sporadic action groups" produce counterexperts and other evidence? Are there perhaps more thorough ways, which science itself might implement, to reflect different schools of thought about the direction that science will take?

Helander also asked what science can do to deserve greater public confidence, if it does indeed rely on this confidence. She suggested that if scientists want to increase public confidence, they must be willing to show their civic courage, to risk their professional reputations when they believe the scientific establishment should be opposed. Toward this end, natural scientists should have a better idea of how society works so that they can better judge the consequences of their research, especially in the areas that border on the social sciences, such as genetic engineering.

Helander's last point was that scientists must always maintain their professional standards; she cited the example she gave in the discussion of David Landes's paper, of the Nowotny study concerning the Austrian referendum that halted the deployment of the country's first nuclear power plant. That example showed how tempting it was for scientists to compromise some of their standards when put in a tough position of advocacy.

In commenting on current attitudes toward science, Stephen Strickland referred to an earlier comment by Robert Lopez that, if people in the fifteenth century were surveyed as to their attitude toward the effect of science on their daily lives, they would have been unaware of and uninterested in the role it played; a similar poll in the eighteenth century, however, might have revealed a highly positive response. "If these are the extremes," Strickland noted, "then we are in a very fortunate position today, at least in the United States." A poll in 1972 indicated that positive attitudes toward science, in various degrees and of various natures, added up to 72 percent, and that this total reached 78 percent in 1976. (He suggested, however, that recent controversies over recombinant DNA and nuclear power might have brought down the total in later years.) In another poll, scientists were held in higher regard, as a profession, than any other group except physicians.

This wide popular support makes intermediary institutions—specifically government—more important, because the government is the institution that must translate popular into financial support. Patterns of governmental assistance, Strickland noted, have varied, sometimes causing concern in certain scientific communities. The biomedical community, for example, was alarmed in the late 1960s when its support leveled off as a result of budgetary constraints; popular support had not abated, but financial support had. This leveling off, Strickland pointed out, has ceased; government support of science has increased, particularly in health and energy research. "So, science is in rather good shape," he concluded, "but if faith and confidence were to become absolutely unqualified, we would be in very serious trouble." A balanced view of science is more necessary than ever before because the complexities of science and technology are greater and their dangers more obvious.

Turning to Brooks's paper in specific, Strickland noted two areas in which he believed Brooks to be overly optimistic. He thought that Brooks failed to distinguish between popular support of basic research and popular concern over certain applications of technology, as for instance in the case of recombinant DNA research. He also felt that Brooks was overly optimistic in judging that public confidence in basic research is unqualified. Polls have shown that there is far greater support for research leading to the solution of readily identifiable social conditions than for research of long-term, open-ended questions. In closing, Strickland warned of excessive optimism that public support for science will continue to flourish in the future; skepticism is likely to grow. "It is extraordinary to me," he said, "that the federal government's investment in cancer research, for example, has continued to grow while progress in terms of the conquest of the disease . . . is very scant." Strickland thought that it seemed to be comparable to Dr. Johnson's observation on a man's getting married for the third time: "the triumph of hope over experience"—and he wondered how long it will last.

Hans Wolfgang Levi addressed the point made by Brooks that dissent within the scientific community was one of the roots of public distrust. First, he asked, does the public demand a consensus among scientists in order to calm its fears and be assured? There has been dissent as long as there has been science, and asking for a consensus among scientists in return for public faith may be asking science to join in an impossible bargain.

Second, Levi stressed that it is important to distinguish between different dissenting scientists. A Nobel prize winner in one branch of science, for example, might not have the specific knowledge necessary to take issue with statements made by scientists involved in a completely different branch. When evaluating dissent among scientists, we must always determine at just what level the dissent is conducted.

Third, Levi maintained that the public, by its own powers, has no way of recognizing dissent; it learns of dissent only by transmission through the mass media. The various media act as amplifiers of such dissent, yet they are also selective in what they convey to the public. Therefore, Levi concluded, Brooks's statement that dissent within the scientific community may lie at the roots of public distrust must be weighed carefully.

In trying to reconcile the positions taken by Brooks on the one side and Weizenbaum on the other, Karl Deutsch referred to the basic choice in science between "errors of the first and errors of the second kind." An error of the first kind would be to accept a new technique as safe and beneficial when it is not. An error of the second kind would be to reject a remedy or procedure which is in fact useful and beneficial. The choice is, since there is going to be a risk of error in any case, a choice between errors; the objective is, of course, to minimize the risk. There is no way of completely doing away with the risk, and an attempt to do so would result in errors of the second kind. Similarly, every foray into the unknown is accompanied by some danger and may lead to errors of the first kind. We must weigh the benefits against the risks, Deutsch said, but we cannot always rely on strict statistical evidence to do so.

In actual practice, he said, scientists choose between the two risks of error in terms of the value they place on the two outcomes. For example, producing pills that help us fall asleep at night or that reduce headaches provides a mild benefit; however, if the pills are thalidomide, then the tremendous risk exists that these will reach pregnant women and cause birth defects. The value is such that, though the risk of actual harm is statistically very small, the results are disastrous; therefore we must not proceed. On the other hand, using a dubious chemical to treat terminal cancer patients, who have little to lose, is very risky, but the value to society of finding a cure overcomes the possibility of negative consequences. It all depends, Deutsch said, on the values.

In the case before us, what are the risks of developing a line of science at full speed as against the risks of slowing science down? In the past man has always applied technology slowly. This application has picked up some speed in our contemporary world, for never before has science been applied so massively as in our time. In the past, relative stability was guaranteed by eight hedges: monarchy, aristocracy, churches, traditions in all areas of life, inequality, poverty, high death rates (social stability today is not possible unless death rates return to their previous levels or birth rates are drastically reduced), and formidable geographic barriers: for example, a famine in India did not concern people in the Rhineland and vice versa. Most of these hedges have undergone sufficient change today to speed up the application of science and technology; thus stability may be impossible for some time to come.

Let us suppose, Deutsch proposed, that some new technology has a 5 percent chance of being completely disastrous. If we do nothing, there is a 20 percent chance of pestilence, war, or famine. On the face of it, it seems that the dangers attached to errors of the first kind are only one-fourth the dangers surrounding errors of the second kind. However, Deutsch added, "It seems to be that science and technology act more as amplifiers rather than modifiers of social values. That is to say, societies tend to get out of science and technology whatever they want of it, not what science and technology offer because of some immanent tendencies." He proposed a first hypothesis that, on the whole, societies dominate and determine the course of their science and technology, rather than the other way around. If a society is warlike, science and technology will give it weapons; if a society is hungry, science and technology will give it more food; if it is wasteful, science and technology will give it new and more ways to be extravagant.

As a second hypothesis, Deutsch proposed that underlying a great deal of the opposition to science and technology is a fundamental distrust of the societies in which people live: "East, West, North, or South." If we believe that societies are, on the whole, predominantly self-destructive, then it is only their impotence that has saved them from self-destruction. "Survival through impotence," he suggested, "has been the secret of mankind until now."

The danger that human impotence can be reduced, Deutsch continued, is now acute. Therefore we have movements saying that small is beautiful, a manifestation of the hope that, if we reduce our reliance on technology, maybe some of us will survive somehow. If, on the other hand, we have the strange belief that human beings can intelligently and responsibly do something about their fate, that democracies are not governed by a solid majority of idiots, and that our elites do not wish to exterminate themselves and their children, then we might say that we can take the risks. The question, Deutsch concluded, comes down to whether we take one line of risk or the other, whether human beings should be given more options or fewer; this question has been a topic of discussion since the Stone Age. Today, however, the problems are more acute because the options are of such greater magnitude.

A discussion followed concerning the role of mass communication, especially television, in providing the information necessary to make the right choices. Scientists, as a caste of high priests, could easily use this device to enhance their influence. Many present felt that this is especially true in the United States, where people with no background in history or politics, motivated solely by profit, have succeeded in destroying the critical faculties of large segments of the viewing public, thereby fostering its gullibility toward possible obfuscations by scientists.

Bruno Fritsch commented briefly on the nature of American television, saying that, while he was in no way its advocate, he did not perceive it to be necessarily a dead end. He pointed out that, with cable television, it is clear that some Americans are willing to pay for quality programming, and suggested that American society is so differentiated and complex that television cannot have a uniformly deadening effect on the public mind. Though business does tend to ruin television, numerous cultural and educational stations do air programs of the highest quality. Business support, indeed, may not always be entirely harmful; after all, business funds science via the support of foundations. There is no need, Fritsch believed, to be as pessimistic about television as some seemed to be.

Fritsch believed that there are, however, other and more deeply-rooted problems in transmitting scientific knowledge to the public: the educational system, for example, is in need of reform. The public in general has not been educated sufficiently to comprehend certain levels of abstraction, and this interferes with its ability to grasp scientific concepts and their applications.

I. Bernard Cohen shared the stated concerns over television, noting that he has never been able to understand how Americans can tolerate advertisements about laxatives and hemorrhoids during their news programs at dinner time. Equally incomprehensible to him was how easily swayed they are by pseudo-medical types in white laboratory coats selling products on the screen. This image harkens back to that of the medieval sorcerer—substituting a lab coat for the conical hat as the badge of authority; the attempt by advertisers to use the prestige of science to sell a product astounded him.

Cohen turned then to Brooks's paper, affirming that one of the reasons why ordinary people tend to have a declining image of scientists is, in part, that they have discovered that science is not infallible and that there is no consensus among scientists. The public's most pressing questions are the ones for which there are no definite answers, probing the borderlines of knowledge; thus, the scientific community can offer no consensus. Questions of the future and of how certain factors will affect certain others in, for example, the environment, would require a universal seer for their answers. Scientists, he stressed, cannot play this role that the public expects of them.

Cohen also commented on the assertion by Brooks that drew Weizenbaum's criticism, namely, that a healthy fundamental research program was necessary for the future prosperity of the American economy. Here, Cohen felt, the important word was *future,* for the practical applications of fundamental research are often unforeseen and unpredictable; in fact, research undertaken toward practical ends would probably not be funded. An example is the development for farmers of hybrid corn, which was discovered in the course of basic research in genetic theory. What the public finds hard to accept, Cohen believed, is that if it wants practical results, it must be willing to permit fundamental research. The most important innovations in, for instance, communication, transportation, medicine, and industry, were all founded on basic research. Convincing the public of this connection might be accomplished through education, but if this is the case, then the mass media must be better equipped for their role in the enlightening process.

Klaus Traube opined that there was probably an underlying consensus to this discussion: that society needs science and scientific progress. While we may discuss the steps of science, or whether we need more or less science, or errors of the first and second kind, we would never question the fundamental necessity to society of some form of science. Fear of science, then, is not a fear of science itself, but of its possible and actual applications. This fact points to the political theory which, in part, science has fostered: namely, that the individual has lost some autonomy as a result of certain structures. Science is the motor of this process, since many of these structures depend on scientific advance.

We have, Traube said, a choice between big science and small science. He qualified the term *small science* by noting that it did not imply that we return science to an earlier status, but that we try to free it from the interference of the economic processes, fed by science, that govern society. There is an interrelationship between the scientific community and the kind of science we have: big science is in the interest of the scientific community because it feeds its research, but small science, he believed, is rather in the interest of society in general. Citing an example from personal experience, he noted that he was formerly one of the managing directors of the nuclear industry in the Federal Republic of Germany, and that, at the time, he believed that nuclear energy was absolutely necessary to serve society's needs. Now, he said, he was equally convinced that we need insulation and solar energy—less big science and more small science. In addition to being less dangerous, small science also has the distinct advantage of being "social" and democratic in the sense that it can be understood and applied by the public.

Max Kaase brought up the topic of the role of political institutions, as opposed to scientific institutions, in the process of determining the direction of science. To illustrate the idea he wanted to develop, he described his experience with the data protection law implemented in the Federal Republic of Germany in 1978. At first, he said, it went unnoticed by the social science community. As its implications began to become clear, however, it became an object of concern. Scientists realized that the new standards could have a tremendous effect on certain methods of social science; the personal interview, for example, could be banned or severely curtailed. So Kaase consulted with an official in Hesse with jurisdiction over the standards, asking him if an internal code of ethics might be a better approach to safeguard against wrongful intrusions into privacy. The official, Kaase related, agonized for a while and finally replied only that whenever professions run into such difficulties, they try to insulate themselves against external political influences by developing self-policing codes. This statement, Kaase thought, was an excellent but frightening indicator of where science as an institution might be headed.

The example provided the basis for two assertions by Kaase. First, he said, science is facing the politicization of its profession, especially social sciences within the university context. The question brought to bear by changing values in society is one of control: will science develop internal institutional control, or will it be controlled externally? His hunch was that it will be the latter. Thus what was needed in such a conference as this one was not so much dialogue between social and natural scientists, but between all scientists and politicians. Second, he said that the conference failed to address the question of where most scientists work. He pointed to Brooks's statement that only 12 percent of scientists function within academia, with the remainder in the employ of private industry or government. Because of their superiority in numbers and their place within our social structures, it is these scientists who are subject to the real potential for control, whereas this conference addressed itself mainly to academic scientists. What the conference was discussing, he suggested, might not be relevant for very long, for he believed that the actions of scientists will soon be determined, not from within their community, but by political decisions concerning society's resources.

Harrison White entered the discussion by stating simply, in the interest of brevity, "Priests breed bishops." To develop this theme, he noted that Harvey Brooks was a bishop, one of several, and that he played the bishop's part of suppressing dissent well, conveying the feeling that there is something immoral about dissent's very existence. Equally important to White's statement, however, was that if priests breed bishops, there must be something in the priests that demands such a role for their bishop.

White then confessed his astonishment over how confused he felt over the proceedings, and admitted that he had tried to simplify them in his own mind by asking himself as each participant spoke, "Do we trust the public?" That question, he felt, split open the discussion. The conclusion he derived, moreover, was that bishops did not seem to think that there was a healthy, realistic collective intelligence in the public. On the other hand, White allowed that the major problems among bishops, priests, and the public emanated from blurred communication.

He derived one other tentative conclusion from his statement. He thought it naive of Deutsch, "our cardinal," to assume in his discussion of errors of the first and second kind, that the public—or anyone else, including scientists—could control the progress or nature of science. Prospects for such a control are hopeless, he maintained.

David Landes sought to include in the discussion a framework that would take into account the international aspects of scientific problems; he noted that the conference had thus far dealt only with the issues within the context of individual nations. Each national unit, he said, had its own political system and its own political objectives, but the problems of science reach beyond these boundaries. For example, in the case of the oil spill resulting from an accident at a Mexican well off the coast of Texas, the United States was a victim, yet Mexico's president announced publicly that Mexico felt no obligation whatsoever to reimburse the United States for the damage. Problems of science and technology,

Landes stressed, affect all nations, yet the conference had not addressed this aspect.

The relationship between science and growth, he further noted, varied between countries, rich and poor. Poor countries, who perceive their need for progress and development to be urgent, seem to be less prudent and less cautious than rich countries, whose needs from science are not as great. In quest of growth, he suggested, poor countries believe that they cannot afford the scruples, related to the use and development of science, so prevalent as of late among the rich countries. The issue of nuclear proliferation, he said, was a case in point, and he wondered whether unilateral disarmament by the United States would be nothing more than a weak ethical gesture in the face of the push for nuclear capabilities by poor nations. Thus, the implications of international politics for science cannot be glossed over.

Deutsch asked to defend his "naiveté," to which White had referred. People's choices do, in fact, influence the course of events, and Deutsch had chosen seven examples of major scientific developments that had been affected by conscious decisions. One was Edison's resolve to construct an industrial research facility, the first of its kind. With the use of this large laboratory, Edison was able to find the electric light sooner than would otherwise have been possible. Second was the negative decision to stop the production of the supersonic transport in the United States; it was halted by scientific opposition in conjunction with public dissent. Third was the abandonment by the United States of the antiballistic missile system, spurred by scientific opposition and political discussion. Fourth was the moon landing; it was a political decision, a planned effort, based on the opinions of several components of society. (Deutsch himself had taken part in a government seminar which asked representatives of finance, academia, technology development, and other institutions how they viewed such a project.) The fifth was the containment of poison gas, a scientific development whose large-scale deployment was mercifully restrained. Sixth was the Manhattan Project, a landmark case of a political decision to speed scientific advance, compressing twenty years' research into five. The seventh example was current research on synthetic fuels and solar energy, stimulated in the United States by presidential and Congressional decisions.

What is important about these decisions, Deutsch said, was their timing. The mere fact that something happens is not always crucial, but that it happens at a certain time can often greatly affect history. Suppose, for instance, Queen Isabella had not financed Columbus; someone would have gone to the Americas, but twenty years later, and the development of the New World would have taken on an entirely different complexion.

Robert Lopez emphasized Platt's earlier point about education, asserting that the status and quality of science education at the elementary school level was a disgrace and that this problem continued through secondary schools to the universities. Schoolchildren, he reminded, grow up to be legislators and politicians. He deplored the fact that no one can become a full professor by only teaching undergraduates. Teaching science, even at the university level, has reached a very low point because scientists themselves do not hold the teaching of their subject in sufficiently high esteem. If science has become unclear to the public, it is because the scientists have failed to communicate properly; there is nothing inherently incommunicable about any science.

Thomas Trautner presented an encapsulated analysis of the issue of genetic engineering from his own point of view as a molecular scientist. He said first that genetic engineering was characterized at once by both a remarkable degree of clarity and a remarkable degree of misunderstanding. The latter stems from the fact that a basic understanding of biology is generally completely lacking, both in the population at large and among those who claim a "higher education," as pointed out by Platt and later Lopez. Such misunderstanding is fostered, as well, by distortions in the mass media, and both factors combine to frustrate a proper understanding of science in general and genetic engineering in particular.

What is genetic engineering? Trautner stated first that it is not an area of research but rather a technique or a method which can be applied to problems of biology. Specifically, genetic engineering is the potential for biologists to take genetic information from genes in one organism and link it to genes in another organism; or, in new combinations, to genes of the same organism introduced into a new organism. Why should we want to do this? First, it gives us the possibility of obtaining enormous numbers of genes, which usually exist only in small quantities. Second, we can reach a better understanding of how genes function through experiments that bring them into new cell environments.

What are the benefits? Trautner explained that, with the help of genetic engineering, we can learn about the structure of genes which are normally unavailable in the quantities necessary for study of their life processes and differentiation structures. There is no such application at the moment, but such goals are contemplated. Such a study might lead to rational understanding of cancer cells; this remains a goal even though cancer is a very minor health problem within the context of the entire world in comparison to, for instance, infectious diseases. Then there is the application problem—and it is here that the Frankenstein image arises—namely, that one can use foreign genes to produce new organisms. However, such an application, Trautner said, is only a possibility, and not a very likely one in his view.

Trautner was concerned that, while fears of genetic engineering were obviously present, no one has offered to analyze these fears. In his estimation, there were four kinds of fear. First, he said, people simply assume that the combination of genes not themselves dangerous will lead to the creation of a dangerous organism. The probability of such an adverse reaction, he said, is not zero—but it is very close to zero: as close to zero as the possibility that two pages from Chaucer inserted into a novel by Oscar Wilde would result in a book describing microwave ovens. Statistically, the evolutionary argument is greatly unfounded.

A second fear is that the combination of a pathogen and a harmless organism would create an organism more dangerous than the original pathogen. He noted in response to this fear that epidemiologists tell us that pathogens introduced into a new material must compete within an environment of stabilized biological equilibrium, and that there is very little chance that pathogens would survive in such a competition. Still, Trautner noted, these two fears are the ones in which reasonable criticism plays a part, and he agreed with Helander that evaluations of these arguments by the scientific community would be sensible and helpful.

Trautner identified the third fear as the belief that genetic engineering would interfere with the identity of an organism, especially as it applies to human behavior. This demands a definition of identity. If what is meant are the social patterns of a person—the capacity to think, to love, to hate—then he did not see where genetics comes in, since only learned information essentially determines social behavior patterns. If, however, hereditary disease is part of a definition of identity, then identity is potentially open to change. If there is any possibility that this technology can be applied to higher organisms, Trautner said, it is in this area.

The fourth fear, he said, harkens back to the old taboo that it is forbidden to tamper with nature. As he noted in a previous discussion, we have been doing this all along, with the breeding of animals, the willful transformation of agriculture, and other such endeavors. Gene exchange does not introduce an essential revision of human intervention in the ways of nature.

In addition, he asked, how has gene technology become so controversial? Scientists have been conducting genetic engineering for many years under self-imposed regulation, and everything has worked well. The important consideration about controversies surrounding recent developments, he said, was timing: they came at a time defined by disillusionment with science in general and the failure of the cancer program in the United States; genetic engineering appeared in the public eye at the peak of the environmental movement. Adding to the distortion was the common fallacy of comparing the problems of genetic engineering to those of nuclear power, as had been implicit in the conference thus far. The dangers of nuclear power, however, are definable and predictable, he said. We can estimate, for instance, what will happen if there is a leak of radiation. The consequences of genetic engineering getting out of hand, however, are unknown, and predictions of both hazard and safety are thus baseless.

The discussion is further complicated, Trautner said, by the fact that it was scientists themselves who first cried wolf, alerting the public to hypothetical dangers with no real evidence. Scientists allowed themselves the luxury of fear, conveyed it to the public, and, now, even if they utilize their methods of experimentation and scientific discourse to shed their initial irrational fears, the public will still be afraid. Because of their expertise, scientists are in a position to reverse their own fears, but the public does not share this position. Misconceptions may be permanently implanted with the public as a result of the early, undocumented pronouncements of fear by scientists.

Finally, Trautner asked, what are the political implications? He mentioned Strickland's example of regulations of genetic research, such as those of the National Institutes of Health, and noted similar regulations in the Federal Republic of Germany. He posed a further question: can there be legislation on a hypothetical problem? The danger is that a bureaucracy will be created that, like all bureaucracies, will never question the justification of its existence; it would ultimately perpetuate the fear of research in order to maintain its own existence. Scientists, he said, are not against regulation. Indeed, they were the first to ask for controls, but these were—and always had to be—self-regulatory controls. An example of such controls is the case of research laboratories dealing with infectious diseases; these laboratories have successfully regulated themselves.

Levi remarked that nuclear scientists were among the first scientists to analyze risks by use of the methods of science and mathematics. This presentation of the relative risks of their field was misperceived by the public; the analyses were based on numerical probabilities, and the public just did not understand the nature of the term *probabilities* in this instance. It only heard that there was a possibility of some danger, and immediately assumed the worst. Next came public pressure to remove the risk entirely. The media, Levi said, have much work to do to alleviate this notion of a risk-free world—such a world simply does not exist. The concept is completely unrealistic in a highly technological world.

Weizenbaum offered some additional remarks "stimulated by Karl 'Cardinal' Deutsch." Deutsch had spoken of science at full speed as compared to slowed-down science, and of the problems entailed by such an evaluation. Using the metaphor of an airplane, Weizenbaum added that it was important to note not only the speed of the plane, but also the direction in which it was going. "Straight down, for example?" he posed.

Weizenbaum noted also that at American universities such as his own MIT, weapons research was no longer conducted. Research is done in such areas of inquiry as "rapidly expanding gases accompanied by large energy exchanges," but no one is conducting research with such titles as "bigger bombs." This raises again the question implied by Gerhard Leminsky in a previous discussion, that if we are to talk about the future control of science, we must ask, "Who controls it now?" If we look at what a scientist does and does not do, apart from the speed and energy with which he does it, it is rather clear who is in the driver's seat. Weizenbaum explained by observing that, at planning meetings such as those in which he participates regularly at MIT, scientists are often conditioned to enter certain fields of research by the fact that government agencies would support certain kinds, but not others.

Weizenbaum observed that Deutsch had implicitly adopted his sports car analogy of a previous discussion, that no one thinks that a sports car is dangerous in and of itself, nor does anyone oppose the manufacture of sports cars. However, no one would give such a vehicle to a fourteen-year-old child. Underlying such an analogy, said Weizenbaum, is a healthy mistrust of society. With some qualifications, this same mistrust underlies the American political system; the nature of the American Constitution is such that it protects the citizens from government by ensuring the government's impotence. Trust in the applications of science might similarly be instilled in the public by demystifying science, by reminding the public of its impotence in several fields, including that of computer science. Weizenbaum related that it was a matter of some comfort to him, for example, that he had access to the FBI computer system and knew it to be incompetent, because he could then be assured of his safety from persecution by virtue of the system's built-in failings.

Claus Offe said that the Ferrari example reminded him of a paragraph in Schumpeter which implied that if we wanted to have faster cars, the decisive thing to do was to develop better brakes. Offe applied this principle to systems analysis, which is broken down into, first, design complexity—the increase in the number of events that can occur within a system—and, second, control complexity, the capacity to select desirable events from the universe of possible events. A balance between the two may have existed in the world Schumpeter described, Offe said, but there is no such balance today in science and its technological applications. Two obstacles, interdependence and irreversibility, stand in the way of control complexity. It is difficult enough today to foresee the effects of technology, let alone control them.

This lack of control complexity becomes important, Offe said, when we view the phenomenon of practical versus theoretical heresy. The standard example of theoretical heresy was Galileo's challenge of an existing world view and cosmology with his own coherent view; this sort of challenge was a threat to authority, to the ruling establishment. The other kind of heresy, practical heresy, challenged orders and traditions, but did not provide a new way of seeing things or legitimizing courses of action. This kind of heresy, Offe said, relates to the practical effects of modern science on the day-to-day activities of citizens.

In contrast to the effect of Cartesian knowledge, which was described in a previous discussion as fostering greater democracy in knowledge, science in our own time has led to production of knowledge that creates an exponential increase in ignorance: the more *some* people know, the less the rest know. For instance, advances in social science have brought us new ideas about bringing up children and keeping them healthy; parents heretofore thought that they knew how to do this in a self-evident way, but now they experience doubt—they feel ignorant. This feeling of ignorance, Offe said, was a root of modern distrust of science.

If there is a gap in control complexity vis-à-vis design complexity, if practical heresies destroy the traditional ways of doing things without providing new world views and orientations, Offe argued that we enter a pathological circle: the control gap, or coping deficit, will have to be filled with alternative forms of practical knowledge, based on standards of goodness and rationality, though it is these very standards that are undermined by the new science. Science has almost rendered individual practical knowledge irrelevant by its advances, yet it provides no control complexity equal to the demands of its design complexity. Offe's posing of this paradox closed this session of the conference.

Chapter 11

Fear of Science—Trust in Science: A Meditation

Joseph Weizenbaum

The two poles of the issue being discussed here are represented by the words "fear" and "trust" respectively. I intend to present a thesis which is rather heavily committed to the fear side of that polarity. I am not, some of you may be surprised to hear, a technological optimist. I intend to argue that thoughtful and responsible people today are correct in rather much more fearing science, on balance, than placing their trust in it. People ought not to believe, for example, that most of our problems can be solved by science. Indeed, I think that human problems are almost never solved; they are sometimes transformed into other problems that are easier to live with (or appear to be so in prospect), and sometimes, after many transformations, the original problems seem to have disappeared. But they were then not solved! I intend also to argue the irrelevance of the otherwise comforting observation that the fear-trust teeter-totter (with respect to science) has, since times long ago, teetered first one way and then the other; that, in other words, wherever we come down today, its earlier state will be restored tomorrow—hence not to worry. History, at least in this respect, is not going to repeat itself. It is much more likely that we are at the end of history. I realize that there is little merit to counseling people to be afraid—though fear can be a life-saving emotion. But, apart from the fact that there is merit in telling simple truths, I will try to earn my keep by giving some recommendations as well.

Before going further, I believe that it is necessary to ask whom we presume the subjects of the fear or trust of science to be. Are we thinking about the scientist, or the well-informed citizen (for example, readers of *Scientific American*), or the proverbial man in the street? Individual scientists may entertain fears of what other scientists are doing but feel perfectly comfortable with their own work. Alternatively, they may have grave doubts about the activities in their own field; I mean not only about the work of their close colleagues, but about their own activities as well. Clearly the merely well-informed citizen, let alone the ordinary man in the street, is in a very different position. Many scientists dismiss the fears of nonscientists as essentially blind fears. I would disagree with this characterization of what troubles lay critics of science. However that may be, it would have to be admitted that if one believes lay critics to be unqualified to fear science, then one ought not to demand of them a blind trust in science.

However, I think that the distinction between the "informed" and the "naive" observer of modern science and its products involves an illusion. The illusion is, first, that practicing scientists foresee and understand the products of their work, as well as the consequences of what they do and produce and, second, that understanding the details of scientific work is necessary to the making of any judgment whatever about the work, its consequences, and so on. To those in the grip of this illusion, the naive observer, lacking the understanding of detail, is disqualified from even having—let alone expressing—fear of science. Of course the corresponding assertion that the naive observer is on the same grounds disqualified from having and expressing trust in science is never put forward.

My own experience with my many colleagues who practice science, and my reading in the relevant literature, lead me to disbelieve that scientists foresee what is to become of their work—very many are not even interested. Besides, it is notoriously difficult for people to make judgments against their own interests. Furthermore, the very way science has to operate, that is, by abstracting and simplifying reality, militates against the asking of certain kinds of critical questions. On the whole, the scientists' instruments and their hypotheses, not their social or political consciences, dictate what questions are to be asked. (I make a distinction here between "conscience" and "interest.") On the question of in what detail something must be understood before it becomes useful to make practical decisions about it, I think it sufficient to observe that most physicists, to pick just one example, use integration without having what a mathematician would regard as the slightest conception of any theory of integration whatsoever. Social scientists seem not to find it necessary to understand the theoretical mathematical foundations of the many statistical methods they use in order to make good and proper use of the methods themselves. On the other hand, I am told that the use by social scientists of powerful statistical methods to put muscles on very weak ideas is not entirely unheard of. I would say that ordinary persons do not have to have studied quantum mechanics or nuclear medicine, for instance, in order to know that exposure to radioactivity is not good for them or that being near exploding nuclear "devices" may be fatal. Moreover, I think that everyone can grasp the difference between a change in one's own body which is not heritable and a change in one's genes which may be inherited by one's children and their offspring and so on, and thus alter the world irreversibly and forever. One need not have even heard of recombinant DNA to understand that. To the contrary, top-rank biologists working with recombinant DNA and believing themselves to be in the race for the next Nobel prize may have the greatest difficulty remembering this difference.

A much more important question than whether this group or that is more or less qualified to have realistic fears of or hopes for science is whether *any* of us remain in a position to understand the products of science or the world that modern science has helped to create. I am thinking, on the one hand, of the very numbers science has forced us to encounter, which are beyond the grasp of the human imagination, and, on the other hand, of the enormous length to which the chain of metaphor, which is the only basis for scientific understanding, has been stretched, and of the consequent impossibility (I would say) of reconnecting modern scientific knowledge with real world experience and especially with wisdom.

When I speak of the numbers science has given us to think about, I recall, for example, that the order of magnitude of the half-life of certain elementary particles is a pico second (10^{-12} seconds) on one extreme, and of others at the opposite extreme a mega year (3×10^{13} seconds). Neither time span is comprehensible to the human mind. I also think of that charming new measure of explosive power modern science has given us, the megaton equivalent of TNT. Phillip Morrison, the MIT physicist, has computed how long an American freight train, each of whose cars is filled to capacity with TNT, would have to be in order that its load of TNT would be one megaton: 30,000 cars long! Who can imagine what it would be like for one ton of TNT to explode—let alone the equivalent of 30,000 cars packed, not in a long train, but in the volume of a suitcase; and to go off, not like a string of firecrackers, but by releasing all its energy in less than a millisecond? The human mind is simply not equipped to deal with concepts that involve numbers of such magnitude. When it comes to computers, it must be said that we now have many computer programs which require that on the order of 1×10^{13} computational events occur faultlessly in order for the program to be executed without error. Never before in the history of science and technology have quite ordinary people—I mean here computer programmers—attempted to control such an enormous number of events, so tightly coupled, with absolute precision. The faith that users of computer systems have that their systems operate properly, that it is even possible in every case to specify what proper operation might mean, is nothing less than touching! Indeed, I believe that the essential incomprehensibility of computer systems provides a kind of core metaphor for much else going on in modern science and technology.

Before I elaborate on the idea of incomprehensible computer systems, let me say a few words about the overly long metaphoric chain I just mentioned in connection with the near impossibility of relating modern scientific understanding to the real world and to human wisdom. I should say first of all that I believe metaphor and analogy to be the fundamental way in which we understand most things. To explain means to make clear what is not yet understood in terms of something that is already understood. Newton understood, within certain limits, the behavior of apples in the gravitational field (as we would say today) of the earth. He then explained—first to himself—that the moon was in certain crucial respects like an apple. I leave aside the question of what the initial bases could be on which we build our understanding of the world as it unfolds before our infant senses except to say that these bases have, in my view, to do with inherent human biological needs and, later, with developing interests. Hence all human understanding is ultimately grounded in the biological structure of human beings. This assertion, if true, has enormous importance (I would say decisive) on the question of limits of artificial, or computer, intelligence. Further discussion of this point would, however, be out of order here.

To say the rest rather quickly, science became increasingly abstract with and following Newton. In particular, it began to lean more and more on mathematical models. As we know, pure mathematics is a game which, if its moves are left uninterpreted, has nothing whatever to do with the real world—no more than does, for example, a particular move in chess have a *meaning* in real world terms. However, mathematical primitives can be associated with real world phenomena, and then the game (say high school algebra) has an interpretation and a meaning in real world terms.

[Here the German word *Bedeutung* serves much better than the English *meaning*. The reason is that the kernel of the former is *deuten* which entails the connotation (*Bedeutung*) of pointing. Hence the *Bedeutung* of something is clarified (to be clear is to be *deutlich*) by pointing to another something.]

Mathematics quickly penetrated to the heart of science; one might well say it became the heart of science and, gradually and with increasing rapidity, what was to be explained changed roles with what was used to explain it. (In what direction does the shadow of a sundial—in the Northern Hemisphere—move? It moves *clockwise*! How was it decided, hundreds of years ago, that the hands of a clock should move in a direction we now call clockwise? By imitating a sundial, of course!) Physical hypotheses, for example, suggested themselves from mathematical peculiarities—physicists faced with, for instance, a singularity in the solution of a system of equations will immediately ask themselves what the physical interpretation (*Bedeutung*) of that phenomenon might be. Progress in non-Euclidian geometries and in tensor calculus laid the foundation for relativity theory. I think it can be safely said that quantum mechanics and all that flowed from it is an extremely strained interpretation of enormously opaque mathematics. Now the point is that modern physicists cannot—do not even attempt to—understand the logical foundations of the mathematical tools they use—that is, they could not, to save their lives, begin with a logical axiomatization of their mathematical systems and then derive, theorem by theorem, the much grosser working tools with which they build the games they then interpret. On the other hand, neither can modern physicists begin with sense data and, taking a few steps at a time, gradually develop the sorts of mathematical models in terms of which, and *only* of which, they understand the world. The connection between the human mind and the real world is thus broken; it no longer exists. One could say, perhaps, that in principle this connection could be reestablished, that there must be a clear path, every step explicit, which connects the axioms of the physicist's mathematics with the readings of his instruments and finally with the physicist's own personal senses. But again, the human mind is not equipped to deal with systems of the enormous size and complexity this undertaking would involve.

What has to be remembered—if spurious counter-arguments are to be avoided—is that a mathematical (or logical) proof is an attempted act of persuasion, a social act. What I am saying here—and I realize that it is outrageous—is that, beyond a certain threshold of complexity, and one much lower than might at first be guessed, no mathematical proof is *compelling!* Again, my point is that the connection between what physics "knows" and the real world has long ago been broken.

I now return to the notion of incomprehensible computer systems. Let me stress first that I use the term *incomprehensible* and not the term *not comprehended*. The latter would imply that it is at least in principle possible to understand, that is, to comprehend, the system in question.

I am making the much stronger statement that there exist in the world today many computer systems (I would say most nontrivial systems doing much of the work of the world are members of this class) which no single person nor any organized team of persons understands in any reasonable sense of the word *to understand*. Very few computer professionals are any longer startled when I make this statement. It is, however, amusing and, as I said earlier, touching, to watch the reaction of eminent social scientists, economists, and so on, when confronted with this idea. About seven years ago Tjalling Koopmans (the economist and Nobel Laureate) and I discussed this problem, and only a few days ago I delivered the news to our fellow conferee Karl Deutsch. The point, I suppose, is that it was news to both these eminent men, who now and then use computers for their own work. Nor could they readily believe it. They (and I think in this respect they are representative of the vast majority of their colleagues and of the generally educated public as well) believe that because computers are *finite* automata, and because computer programs are designed by people, and because, with the advance in the art of electronics, hardware errors as such can be effectively discounted, computers no longer "make mistakes." If a computer system nevertheless behaves in unintended ways, then an error must have been made, and the mistake must be one caused by a human, a programmer; by a step whose logic escapes me, it must be capable of being found and corrected.

Let me grant, for the sake of argument only, that electronic devices as such are today error-free. The rest of the above reasoning still does not follow. Perhaps the most important hole in the chain of reasoning lies in the mistaken assumption that modern computer systems *have* authors and that they *are* designed. In fact modern computer systems normally *evolve* in a series of disconnected, not to say chaotic, steps. Each major step is likely to be implemented by a team of programmers different from and out of touch with the teams who implemented the previous steps. What emerges is a system which may approximate the behavior that someone in some now-distant past imagined for it—but even then only quite vaguely. The rules for its actual input-output behavior are nowhere recorded and cannot be derived by systematic observation of the system —especially not without making inductions, that is, without outrunning what the evidence of the experimental use of the system permits one to deduce. Since knowledge of the system's input-output rules is required in order to use the system at all, such knowledge has to be obtained somehow. Typically this is done by treating the system as a living organism and inferring its behavioral rules by essentially "psychological" methods. The penalty for employing this tactic is, of course, that neither the behavior of the system, nor the system as such, can be modified without endangering its "life." In the absence of a real understanding of its logic, any modification of the system may induce unknown, possibly very dangerous, side effects which may not come to light until their practical effects have long since propagated into the real world. In the end then, the user adapts his behavior to the divined behavior of the system, rather than the other way around. A quite similar argument can be made for nuclear power reactors.

The importance of the facts I have just recited to the question of fear or trust in science should be obvious. How can we place our trust in things which even their architects and their journeymen cannot comprehend? When we look at the power of the things science has given us, especially the power to do us all in, then the question of trust quickly turns to one of fear.

This is, of course, precisely the point where the historical analyses take over and the talk of the teeter-totter begins. Furthermore, the assertion which every Cassandra has also pronounced, namely, that "this time it is really different," is laughed out of contention. Well, prepare to laugh: *This time it is different!*

To illustrate that we are living in times today which differ *cosmically* from those just a hundred (one could say fifty) years ago, imagine that a catastrophe had wiped out all human life on this earth a hundred years ago. Suppose that then a time comes, perhaps a thousand million years from the time of the event, when all traces that human beings, an intelligent race, had ever existed on this earth have vanished—even the pyramids have by then crumbled into dust. A newly emerging intelligent race would simply find no evidence that we were ever here, no matter how fine their instruments or how great their skill. Were, however, the same catastrophe to occur today, then at that far distant time, a new race of people *would* find traces of materials which they would know nature could not have created—hence they would know that once we were here and that we had learned to light the atomic furnace. Beyond that, we have sent vehicles into space whose limits of survivability are so distant that, from our point of view, we must say that they will last nearly forever. The point is that mankind has, within the last microsecond on the scale of the history of the universe, grasped godlike powers for itself: on the one hand, the power to put its sign indelibly on the universe, and, on the other, the power to commit species suicide. It is no longer a question of the balance tipping one way for a while and another way for another little while. The historical analogy breaks down because the world which could be the only anchor for the analogy has disappeared. Our institutions have reached escape velocity.

My own view is—recall I promised some recommendations—that we have become intoxicated by our science and our technology. (I speak here

primarily of the Western world.) The signs of intoxication are a stupe-
fication of the senses to the point where physical and mental control can
no longer be exercised, and an accompanying excitement or elation to the
point of frenzied enthusiasm. Anyone who has listened to the SALT II
debates before the United States Senate Armed Services Committee,
especially the testimony of Dr. Kissinger, can hardly escape the conclusion that we have lost mental control of our affairs. (The wife of an
American President *christens*—just think of what that word means—a
submarine; her husband boasts to the United States Senate that its
missiles *alone* could destroy life [though not necessarily property] in the
whole of the Soviet Union. Who can fail to be impressed that we operate
in a routine frenzy of excitement and a normalcy of madness?)

The menace which has invaded us, as seen from the perspective of an
information scientist, is fundamentally abstraction. To abstract means
to draw away from. Science, in order to function at all, must practice
abstraction in that it must necessarily simplify and deal with idealized
models—in other words, draw away from reality. Granted, science,
idealization, and abstraction are good and useful, in proper dosages,
though even then only when compounded by wisdom gained from many
other perspectives. However, we began a long time ago to confuse the
abstract with the real and then to forget how to make the distinction at
all. Our increasing loss of contact with reality is illustrated by, in
addition to the examples I have already given, the march of abstraction
with respect to the products of human labor and of human labor itself:
people once traded their labor directly for goods. Then money became an
abstract quantification of human labor. Then checks and other financial
instruments became abstractions for money. Now we approach the so-
called cashless society, in which electrons racing around computers out of
reach of human senses become abstractions for financial instruments. An
observer from another planet will see people laboring in order to optimize
the paths of electron streams flowing on their behalf, in computers
unseen and incomprehensible.

Perhaps the most pervasive evidence of the phenomenon I am trying to
describe is our substitution of peoples' images for their real persons. This
applies not only to individuals, for example to candidates for political
offices and to other so-called celebrities, but to entire populations.
America's war on Vietnam was fought largely to impress various "audiences"—the word comes straight from the Pentagon and from our
State Department. The image of the United States was at stake, not the
lives of real people. Things are treated no differently: an American
Secretary of "Defense" (I put that last word in quotes because it is an
Orwellian euphemism), Melvin Laird, once asserted that American
hydrogen bombs are not weapons of destruction but "bargaining chips"

of which America had to have a great many so that she could "disarm from a position of strength." This is a logical extension to world scale of the already well-established military principle that villages often have to be destroyed in order that they may be saved.

At still another and deeper level, the reduction of authenticity to imagery serves to stupefy the collective consciousness of the people—just as intoxicated minds are stupefied—so as to render unnoticeable much more subtle manipulations of reality, even more profound drawings away from reality. I have in mind the corruption of everyday language, hence of the creative imaginations of speakers of everyday language, through the illegitimate raising of science-based metaphors to the status of common sense truths. It is, for example, now commonplace to hear of people being programmed. In this way does the abstract notion of an abstract machine—and one that fascinates the general public almost to the point of hypnotism—become symbolic of human beings. Once we accept that, metaphorically, human beings are machines, merely symbol manipulators and information processors, then the final step, namely, the deliberate initiation of a program to alter the course of biological evolution in such a way that the human species is replaced by "silicon-based intelligence," can be announced by the most eminent scientists— for example, by Dr. Robert Jastrow, head of NASA's Goddard Space Flight Center—without alerting anyone that what is being talked about is not merely the death of the human species, not merely the death of God, but the murder of God!

It seems to me obvious that what is now needed is an energetic program—at least for the Western world—of technological detoxification. I would say that such a program should involve at least the following steps:

1. We must admit that we are intoxicated with our science and technology in roughly the way I have here sketched: that, in other words, we are deeply committed to a Faustian bargain which is rapidly killing us spiritually and will soon kill us all physically.
2. We must muster the courage and the will to insist that we can recover. My own view is that we cannot recover except with the help of a miracle. I do not consider this a pronouncement of hopelessness, as I will show later.
3. We must make the decision to withdraw from our addiction. We must, for example, decide affirmatively to halt the Orwellian corruption of our languages; we must decide that life is better than death. We teachers must decide to teach, mainly by our own example, that authentic living, not life by abstract formulas, is possible.

4. We must strive to live first this very day, then the next day, and so on, one day at a time. We must wrest ourselves from the framework of abstractions we have erected, and recreate a world peopled by genuine human beings, not by images, a world in which the value of word and deed inheres in word and deed itself, not in the engineered applause of some imagined audience.

I am aware that these four prescriptions are very general. I think that they may serve as guides to actions, but lest I be misunderstood, permit me to make some very specific proposals.

On what appears at first glance to be a not very serious plane, but is, in fact, crucial to the mental health of our civilization, I would say that we in the West must stop the insane consumerism whose principal manifestation is the invention of products followed by the creation of a need for them. There are thousands of relatively trivial examples of what I am talking about. How people visiting America for the first time must wonder that we have such things as electric carving knives and electric combs!

As a computer scientist, however, I am keenly aware of the virtually paradigmatic role the computer plays in this respect. The computer in our society is in large part a solution in search of problems. The mentality which breeds and nourishes this condition is precisely the same which converts human and political problems to technical problems and then proposes technical solutions. One effect of this conversion, and it is not always unintended, is that it distracts attention from real conflicts and from real clashes of interests. There is an almost worldwide initiative, for example, to introduce computers into schools. However, once the abstract characterization of the educational process in information-theoretical terms is ripped from in front of one's eyes, once the reality of what is happening in, for example, American secondary schools, is exposed, then it becomes clear that the problems facing educators everywhere are political, financial, and spiritual. Not least among the causes of these problems, in America at any rate, is that such a large fraction of our energy and wealth is invested in killing-machines. What I propose is that we begin to learn to assess our situation from many different perspectives and without prior commitment to technological fixes. If our assessments reveal problems of a technological nature, there will be plenty of opportunity to bring our technology to bear.

In the last analysis, nothing is a more concrete nor a more dangerous manifestation of our confusion of the abstract with the real, and of the madness of the logic which then leads to murderous policy consequences, than the ongoing international arms race. It is perhaps merely a grim joke—but I think it points to a tragic reality—that the foundation of the so-called defense policy of the NATO alliance is officially called MAD, an acronym for "mutual assured destruction." The first and most urgent step in the detoxification process therefore ought to be our withdrawal from the myth that ever more numerous and ever more powerful weapons of mass destruction offer any security whatever to the peoples of the world. Let me very explicitly declare my own interest: I argue for worldwide and total nuclear disarmament and, as a citizen of the United States, my personal position is that my country ought to begin that process unilaterally if necessary. However, as a first step and in line with the prescription that, in the initial stages of detoxification one attempts to master one day at a time, I plead here only for the small step that we stop adding to our already enormous nuclear arsenal, that we begin to reverse, not merely to "control," the arms race. Moreover, I want it to be clear, as well, that while I believe the destruction of the world's nuclear stockpile to be a practical necessity, I think it even more important that people understand and press it as a profoundly moral matter.

Finally, I feel compelled to say to you that my confidence that we can pull out from the fearful crisis which imperialistic scientific rationality and instrumental reason, among other things, has produced, is very small indeed. I think we are now all passengers on the Titanic. Our instruments tell us where the icebergs are and our computers warn us that no maneuver we can perform can prevent the fatal collision. My impression, gained from spending many years with generations of bright young students and also with my own children, is that the youth of the world knows this. We have put them into a position where they must confront their own death virtually before they have begun to live. Is there no chance to survive? I believe there is exactly one: a miracle is required to save us. Are there still miracles? Oh yes—when Ms. Corrigan and Ms. Williams stopped the fighting in Ireland for a time, the sort of miracle I am thinking of took place.

Is it then sufficient that we all sit back and wait for the miracle to happen? Decidedly no! We must all do what we can to prepare the soil for the miracle's growth—or at least we must not do things which poison the soil from which it might spring. I would say that to spend one's talent in the development of machines whose sole function is to kill people poisons the soil—as does the doing of work one knows to be trivial or merely self-serving. I would say that telling the truth and being good to one another nourishes the soil. A major part of the truth which needs urgently to be told is that modern science does not have, nor can it possibly have, the whole truth about anything. When scientists are clothed with the mantle of omniscience, either by others or by pretensions of their own, they and their sciences become fearfully dangerous. It is the special responsibility of intellectuals, it seems to me, to expose this form of imperialism, this attempt to dominate a territory not legitimately science's own. Imperialisms of all forms plead for trust, but they are all to be feared and resisted.

Chapter 12

Discussion of
"Fear of Science—Trust in
Science"

Summarized by Andrei S. Markovits and Karl W. Deutsch

Concluding the presentation of his paper, Joseph Weizenbaum stressed the special responsibility that intellectuals have to expose what he called "scientific imperialism." Scientists should not, he emphasized, act as if they are clothed in mantles of omniscience. Moreover, they should not attempt to dominate areas not legitimately their own. Above all, they should refrain from pontificating truth, a scientifically unsound and morally reprehensible behavior. He illustrated his point with an anecdote from the days of the 1965 Watts riots, a time of racial, urban, and collegiate unrest. A mass meeting attended by students, faculty, and administration was held at MIT.

> Our Dean of Engineering, who is well-known to some people here, stood up and said, "Why didn't they tell us that they had a problem in the cities? We at MIT, with our techniques and knowledge, could have fixed it!" That is the kind of imperialism I think needs to be corrected!

Jean Stoetzel began the discussion with criticism of Weizenbaum's view of abstractions. He perceived that Weizenbaum held abstraction to be something "bad"; Stoetzel stressed, to the contrary, that abstraction is fundamental to understanding and that it is of great relevance to the topics under discussion. Abstraction, he argued, is the *essence* of science.

The problem for the physicist, for example, is to give physical and cosmological interpretation to abstractions; the task of science is the construction of abstract models. Kant was awakened from his dogmatic sleep by reading Hume. Hume said that the essence of human reasoning is habit. "The real scientists," Stoetzel maintained, "have been trained by their own habit to believe they understand: understand in the sense of Hume, namely that human reasoning, or habit, is part of a chain of causes and effects." However, scientific training is necessary to realize this sense of understanding and reasoning, Stoetzel said, and this fact means that the lay public cannot follow with the scientists in their comprehension of abstractions. Stoetzel chose to differ from Harrison White, who earlier in the conference had said that there are two classes of people: those who trust science and those who do not. Stoetzel claimed membership in a third category: "We just do not care, and we still go on."

Ours is not the first time in which man has been faced with irrational phenomena, Stoetzel said. The mathematics of negative numbers, zero, and irrational numbers at first had no meaning to the lay public. What, after all, did negative-two cows mean to the medieval farmer? Stoetzel argued that the same lack of conception holds true today. Concepts which are elementary for scientists of the present day remain incomprehensible to the lay public. He concluded by saying that the public should be given as much knowledge as it can understand, but that it should not be given the chance to discuss things that it is not "used to," in the sense of Hume.

Weizenbaum heartily agreed that scientists must "draw away from reality"; that is what is meant by abstraction. In science, abstractions can be good and useful in proper doses. "But what I complain of," he continued, "is not abstraction as such, but rather the confusion between what is abstract and what is real." Weizenbaum objected to such abstractions as when the State Department explained that the Vietnam War was fought for "audiences," or when Nixon's Secretary of Defense, Melvin Laird, called H-bombs not weapons of destruction but "bargaining chips" (of which America must have a great many in order to negotiate from a position of strength)—implying that what transpires between the United States and the Soviet Union is nothing more than a poker game. In these instances, Weizenbaum argued, there is a confusion between the abstract and the real, a confusion that could be fatal to us all.

In reply, Stoetzel posed the question, "Would you agree that as long as the public understands something which is supposed to be scientific, it is not scientific?" Weizenbaum answered that he would most certainly not.

Continuing with the theme of public versus scientific understanding, Karl Deutsch proposed three definitions of understanding which he had developed in his earlier work. First, to understand is to describe something unfamiliar in terms of analogies with something familiar, such that behavior and habits can be transferred from the familiar to the unfamiliar. Weizenbaum's examples, however, showed how familiar analogies could be grossly misinterpreted. "Hydrogen bombs do not resemble poker chips very closely," Deutsch concurred, "and they would make poker games rather unhealthy if they did." He commented that there has been a great deal of misuse of metaphor in political propaganda.

Deutsch's second definition of understanding was to make a step-by-step structural model of a process, such that we believe that, as we manipulate the model and it produces certain outcomes, the model will predict, or parallel fairly closely, the outcomes of the process. This model process, however, has its limitations. In his *Logic of Modern Physics*, Bridgman stated that models have limited domains of validity. Models, he maintained, are always composed of components, the real inner workings of which we do not really understand but rather take for granted. Bridgman also said that, in Newton's day, gravitation was an unfamiliar incomprehensibility; by the end of the nineteenth century, however, it had become a familiar incomprehensibility.

The third definition of understanding is used more often in the social sciences. This kind of understanding involves not only using our rational faculties of step-by-step reasoning according to certain standardized methods such as logic or mathematics, but also calling upon our emotions to try to find out how a certain actor may have felt. Wilhelm Dilthey and Max Weber were pioneers in this type of *verstehen*, or understanding. (In the computer example, this kind of understanding can be put aside.)

Deutsch interpreted Weizenbaum to say that if we could not mentally build a reliable step-by-step model of the process of a big computer, we could not say what it would do. He judged that Weizenbaum's style of argument was deterministic, not probabilistic. Deutsch argued, however, that in a "cost-benefit analysis" of computers, a reasonable course of action would be that we should use them; most of the time, they give us information which is useful. Of course, given the fact that we do not totally understand the process of big computers, we must acknowledge that at some point something may go wrong. This does not mean, Deutsch believed, that the bad outweighs the good. Quite to the contrary, the benefits we have hitherto derived from computers far outweigh the costs.

Deutsch argued the point further by employing another example: our faith in airplane flight. We fly on jets in spite of the fact that we do not know all the workings of all the parts of the plane, or enough solid state physics to know about metal fatigue and tiny cracks which may develop over time. When discussing a computer, we should not ask whether the big complicated system is comprehensible to us only in a very gross approximation of its fine details, but rather whether it works well and often enough that it will do us less damage to use than not to. Deutsch drew an analogy between his point and the case of fire and primitive man. Although "this humanoid" could not understand fire in scientific terms, and although fire may sometimes have brought calamity upon him by burning him or the furnishings of his cave, he could not have survived the last Ice Age without it. Deutsch added, however, that we should differentiate between macrotechnological projects on the whole and those specifically connected with what we call defense. Weizenbaum's paper supported the proposition that a policy of deterrence is a gun that goes off in two directions.

Deutsch concluded his remarks by stating that our guiding criterion should always be the following question: "Do we on the whole know enough about a scientific/technological process that we can say with reasonable certainty that its implementation will do us more good than harm, or that by not undertaking its development, we will, in turn, increase problems at the expense of the general improvement in social life?"

Weizenbaum agreed with Deutsch's propositions about kinds of understanding, but he suggested that we should not forget that there are different *levels* of understanding. To understand *King Lear*, for example, is more than to be able to tell what happened. "Most of my colleagues think that computers can understand *King Lear* in the same way that I, having four daughters, can understand *King Lear*," he explained. "And at my age!" Weizenbaum warned against accepting the "touching faith" in computers that their users sometimes have. He saw the computer in a different light than Deutsch presented it in his cost-benefit analysis; Weizenbaum was not willing to take the risk of assuming that nothing will go wrong with computers. There is a crucial difference, he argued, between taking the risk on a trans-Atlantic flight and taking the risk on a computer. The difference is this: once you get off the plane, the risk is zero; the flight has a beginning and an end. With computers, this is not the case; there are some computers to which we are nearly "irreversibly tied": we use them all the time. The risk that something may go wrong is *always* there, since we use them continuously.

The major point is not even that at some stage an "error" that the computer makes may have disastrous consequences; the probability of this is indeed very low. The major point, rather, is that "we are now forced in many instances to treat computers as if they were living things. We can no longer adapt computers to our wishes, but rather we must adapt our behavior to their 'desires,' so to speak: that is, to the way we perceive them to operate." This "we," Weizenbaum said, includes both the computer scientist and the general public.

Weizenbaum further illustrated his point by noting that he knew a high official at the Defense Department who was unaware of any computer system in the department without errors in its program. The people who deal with computers, therefore, have to adapt to the errors which lie within the system. This offers a very different perspective on computers than that suggested earlier by Deutsch's cost-benefit analysis, in which a computer is "worth it" because it works most of the time and is useful even though it may make a mistake at some point.

The people who deal with computers are compelled to treat them as living things, as natural. "But," Weizenbaum reminded, "computers are *not* natural." Certainly this is not a unique circumstance. Man has dealt before with other forces he could not always control, such as fire, foreign enemies, and rivers which burst dams. The missing point is that if we are to live with computers, inherent errors and all, then we ought to *decide* to live with them and their inherent risks. This is not now the case, since there are still many people who believe that computers are comprehended, comprehensible, and under the control of the people who use them.

Weizenbaum then argued passionately that more controls are needed not just for computers and their programmers, but also for nuclear energy plants and their engineers as well. The following describes both situations: the programmers deal with unexpected anomalies; they compound the problems while trying to solve them. The instrumentation systems are inadequate; we do not know what the computer or nuclear reactor is doing. The regulations for monitoring both computers and nuclear reactors are also inadequate. If we are to build computer systems that will have more complex capabilities—and not just let them happen without controls, almost by accident—then we must *decide* to build them. The decisions should not be based on vague generalities, but rather on concrete, case-by-case cost-benefit analyses.

Harrison White argued that the characteristics and problems that Weizenbaum ascribed to the computer world and the nuclear energy industry apply to all large formal organizations. He suggested that there is nothing unique to computers that fosters this situation.

Bruno Fritsch raised some points that compelled him to "profoundly disagree with Weizenbaum." He stated that he could not believe that the world is going to end deterministically. (Weizenbaum interjected that he did not believe this either.) "The world will go on," Fritsch said. "The question is: how?" Fritsch also returned to the issue of mathematical proofs which Weizenbaum raised in his paper. "Why is a mathematical proof a social act?" he asked. This question sparked a heated discussion among conference participants over Weizenbaum's assertion that a mathematical proof is an act of social persuasion.

Weizenbaum explained himself by using the example of Goedel's proof. "How many people in the world today," he asked, "have carefully combed Goedel's proof and decided, 'Well, he's right'?" There are certainly a small number of people who have done this, but even those few, Weizenbaum argued, have not reduced each step to its binary terms. There is a consensus in the scientific and mathematical community that certain theorems are established and may be used in proofs such as Goedel's. Weizenbaum viewed this agreement among mathematicians as a social phenomenon.

A long and abstract discussion about axioms followed, concluded by White, who emphasized that to derive a mathematical proof is very difficult. The derivations, he said, "are remarkably undefined and slippery, because they are so long, that in all kinds of subtle ways, little conventions and *cultural* habits creep in." These conventions have arisen to simplify impossibly complicated situations. This process is used in the experimental sciences as well, White argued, and the problem with this is that the conventions may allow something to be proven true which should not be the case *prima facie*. Perhaps in the course of time these conventions will prove false.

Klaus Traube turned to the discussion of Weizenbaum's point that the scientific world of today is fundamentally more complex than that of one or two generations ago. Evidence of this includes not only computers, but also genetic and nuclear research. He pointed out that the change did not occur because of a lack of understanding: whether we or previous generations understand gravity is not relevant. Gravity has always been there; it did not change the world when someone tried to express this force in the form of physical law or scientific formula. What *did* change the world was the production of new machines, based on such scientific formulae, which definitely went beyond immediate human experience.

Traube continued his remarks by stating that we are now being overfed with new information. Like Claus Offe in a previous discussion, Traube maintained that exponential growth of knowledge creates ignorance because we cannot possibly comprehend or digest the magnitude of the complexity of all this new knowledge; there is simply too much of it. He illustrated his point by noting that in the ancient world there were approximately 100 different professions. Historians generally agree that an ordinary person could more or less understand what each of these professions entailed. In the United States today, however, there are 40,000 professions listed. How can we possibly comprehend this? He suggested that we cannot. This quantitative growth in complexity also meant a qualitative change in the development of access to knowledge and information. It could easily lead to the growing ignorance of the vast majority of the population, as against an ever-shrinking but increasingly powerful technocratic elite. Stoetzel had suggested earlier that people be given only a certain amount of knowledge. Traube raised the question of who was to decide what the public should know. More importantly, he wondered who has the right to make these decisions. He concluded by saying that people have a right to understand what is going on around them, and that in today's world, they cannot. This situation represents a fundamental threat to democratic society.

Peter Weingart attempted to evaluate the assumptions made during the discussion about the public. He said that the public judges in terms of its own experience. In making scientific judgments, the public differentiates very knowledgeably among the various scientific fields and their respective values to society. Medical research is always held in high esteem, for example, irrelevant of its apparent successes or failures, because people can relate to the medical world; they have had personal contact with it. When contrasted to medicine, high-energy physics, for instance, is not well-regarded by the public because it is not known.

Weingart believed that the public's view of science, though ambivalent, is well informed on the whole. On the one hand, there is a general trust in science. On the other hand, however, there is specific distrust for certain projects or developments. Such distrust comes from perceived dangers which are rooted in personal experience. The public does not judge in terms of any kind of scientific understanding; this, however, does not make its conclusions less valid. When personal experience is no longer the frame of reference, then the public turns to the experts. The situation in the scientific world today is such that scientists often disagree among themselves on certain issues, such as recombinant DNA or nuclear power. While this may be attributable to the complexities and dangers of various scientific projects, and is neither surprising nor reprehensible in and of itself, it leaves the lay public in a very strange position. It no longer has any authority to depend on, and consequently, anomic behavior and attitudes result. Ambivalence, therefore, is the only logical outcome of this process.

What does this imply? The scientific community, said Weingart, must make assessments of the dangers of technology. These assessments are not based on real experience—no scientist has ever seen a nuclear reactor meltdown, for example—but rather on models. Although the public understands very little about calculating probabilities, it does know that scientists are making rash assessments based not on facts, but rather on abstractions. To most people, this is very alarming.

In recent history, science reached a stage at which it was able to promise that it could make good predictions and develop technologies that would lead to a better life. The vast majority of the public believes this, Weingart reported, and is not ready to accept what scientists are now saying, which is that there are dangers in our technologies that may indeed be catastrophic. Technologies are now being assessed before they are implemented; this was not the case in the past. Hazards of automobiles and trains—for example, the fact that they may crash—were realized only after the technology had been implemented. "We have reached a state of reflexivity where science and technologies are being assessed before they are implemented," Weingart continued, "and that is why the scientific and technological communities are surprised. They are being caught in the act, the act of evaluating themselves." Weingart concluded his comments by asking, "Do we want to give up the public for the sake of scientific progress, whether the public believes in it or not?" He felt this to be an important normative question. If we as scientists and intellectuals answer yes, Weingart added, we are in trouble.

In his concluding remarks, Weizenbaum said that we have no reason to fear for lack of adventure in science; there are more than enough people who are willing to sacrifice the human population for science's sake. Weizenbaum turned once more to the idea of models. If we want to be able to control a technological system at a certain level, we must have a model which is of the level of complexity that we are able to understand. The situation now is that we have technological systems that we can voluntarily build or not build. Because these systems have not yet been designed, it is very problematic to talk about control complexity commensurate with design complexity. At this point the discussion was concluded to allow time for the next presentation.

Chapter 13

Eight Major Evolutionary Jumps Today

John Platt

Many similes have been used to dramatize the revolutionary nature of our technological jumps since World War II. Nuclear power and nuclear weapons have been compared to the discoveries of fire and gunpowder. Television, with its wide dissemination of information, has been compared to the invention of printing. Our first steps toward living and working in space, which Gerard O'Neill has called "the high frontier," are being compared to the opening up of the American frontier in the last three centuries.

However, I think we can make larger comparisons that are even more impressive. Several of our jumps would seem to be as great as the most extensive evolutionary developments in the whole previous history of life on earth, to the degree that such comparisons are possible. The intersection of several jumps on such a scale simultaneously in this generation then becomes an overwhelming fact. We see that it may be a singular epoch not merely on the scale of millions of years, but of hundreds of millions of years. Such a realization cannot help but change our whole view of the human situation today. Our immediate future and our long-run future, under the pressure of these sudden great developments, will surely be different from anything that has ever gone before.

Table 1. A Classification of Major Evolutionary Jumps

Functional Areas	Eras					
	Early Life (4000 million years BP[a])	Multicellular (1000 million years BP)	Early Human (3 million years BP)	Post-glacial (10,000 years BP)	Modern (600 years BP)	Present Transformation (40 years BP to present)
Genetic mixing and control	SEX-CROSSING	Migration		DOMESTICATION AND BREEDING	DISEASE CONTROL, CONTRACEPTION	MOLECULAR BIOLOGY, RECOMBINANT DNA
Energy conversion	PHOTO-SYNTHESIS	PLANT-EATING	FIRE	AGRICULTURE wind, hydro	COAL STEAM, ELECTRICITY	NUCLEAR FISSION, (FUSION) SOLAR ELEC-TRIC, (SPACE POWER)
Encapsulation and habitats	CELLS ocean niches	Shell, skin, bark LAND	Clothes all climates	CITIES all continents	West "frontier"	SPACE CAPSULES, (SETTLEMENTS) Arctic, ocean
Methods of travel	Drift	Fins, feet, wings	Boats	Horses, WHEELS, SHIPS	RAILROAD, AUTO, AIRPLANE	Jet, ROCKET
Tools and weapons	Chemical	Teeth, claws	TOOLS, WEAPONS	METAL	MACHINES, GUNS EXPLOSIVES	AUTOMATION, ROCKETS, NUCLEAR WEAPONS
Detection and signaling	Chemical	HEARING, VISION, echolocation	SPEECH	WRITING	PRINTING telephone, radio	ELECTROMAGNETIC SPECTRUM—RADAR, Laser, TELEVISION
Problem solving and storage	DNA CHAINS	NERVOUS SYSTEM AND BRAINS	Oral memory, prediction	MATH, SCIENCE, LOGIC	SCIENCE AND TECHNOLOGY	ELECTRONIC DATA PROCESSING FEEDBACK CONTROL
Mechanisms of change	Accident and SELECTION	Foresight, REINFORCE-MENT	THOUGHT	INVENTION	RESEARCH AND DEVELOPMENT	SYSTEMS ANALYSIS AND DESIGN PROJECTS

[a]Before present.

COMPARISON WITH EARLIER
EVOLUTIONARY JUMPS

Table 1 is an attempt to make some meaningful comparisons between our recent developments and earlier evolutionary jumps. The columns indicate broad historical epochs, dated by their time before the present (BP). The rows represent different categories or functional areas. They were chosen so as to encompass eight recent jumps as given in the last column, but were made broad enough to include the earlier developments as well, so that a comparison is possible. In some of the rows it might even be possible to make numerical order-of-magnitude comparisons of the changes in speed or complexity or other characteristics of the different jumps, but qualitative comparisons are all that are intended here.

Row 1, Genetic Mixing and Control. This category was intended to include the successive great advances in methods of biological evolution. The first of these was probably the development of sexual crossing between male and female cells of the same species. This must go back over 3,000 million years, since fossil bacteria found in the rocks of that period seem to be very similar to the bacteria that show sexual crossing in laboratory studies today. Under natural selection, however, this method took several million years to create a new species. With the coming of human selection, in the domestication and breeding of plants and animals some 8,000 years ago, the time to create a new species could be speeded up to a few hundred years or less.

In the last century, the germ theory of disease and new methods of contraception have radically changed our control over biological reproduction. However, many biologists would now say that all these earlier advances are overshadowed by the discovery of molecular biology and recombinant DNA methods. These methods seem to permit the insertion of genes from any species into any other species—viruses, bacteria, fungi, plants, or animals—and thereby to create millions of new species overnight. Human insulin is now being made by bacteria in laboratory flasks. We have become the conscious agents of evolution, and in terms of the million-year or the billion-year future, this would appear to be the biggest evolutionary jump in biological history.

Row 2, Energy Conversion. Our newly developed photovoltaic cells for the direct conversion of solar energy to electricity are probably more important for the long-run future than coal or oil or nuclear fission (which is also based on limited fossil supplies). Nuclear fusion power, using hydrogen from the oceans, could also be of major importance if it proves feasible in the next few years, as is now expected. These are major new evolutionary jumps, certainly comparable in long-run importance to photosynthesis. Until now, photosynthesis, which developed in blue-green algae some 2,000 million years ago, according to current archaeological dating, has been essentially the only energy source for all life. The human development of fire, agriculture, and the age of coal and steam, are only more effective ways of using it. However, direct solar electric cells are far more efficient than photosynthesis. They eliminate the need for harvesting and burning wood and coal. Furthermore, in the form of solar satellite power stations, if these prove feasible, solar electric cells could provide continuous and inexhaustible power from sunlight for every nation on earth, far beyond our present rates of consumption.

Row 3, Encapsulation and Habitats. Encapsulation is needed for any new habitat. Our jump into the new medium of space today can be compared to the coming ashore of the land animals, as Wernher von Braun often emphasized. This occurred 500 million years ago, but we still carry encapsulated within us the ancient salty ocean; furthermore, we must still coat our eyeballs with ocean water every time we blink. Thus today the exploration of space has been made possible by space suits and space capsules carrying terrestrial air, water, and food. Recently, Gerard O'Neill has shown that with present technology and with not too great a cost, we could create space settlements for thousands or millions of people. They could be in great cylinders or wheels that rotate to give artificial gravity, and they could be large enough to have a blue sky, clouds, lakes, and even high mountains to climb at the ends of the cylinders. He envisions a rapid multiplication of such settlements, as with the migration to the New World, with millions or billions of people in space, and with abundant resources of energy from the sun and materials mined from the moon or asteroidal rocks. If this proves feasible, even on a slow time scale, it could be far more important for the long-term future than the opening up of a new continent or even the evolution of the land plants and animals.

Row 4, Methods of Travel. The most dramatic previous changes in ease and range of travel probably came with the invention of wheels and seagoing ships some 5,000 years ago. The jump of rockets into the new medium of space is surely as great as, or greater than, the invention of ships as measured by almost every parameter.

Row 5, Tools and Weapons. These are grouped together because in earlier times they were often the same, from teeth or fire to explosives. Today the most brilliant development in tools or machines is in auto-mation, and the greatest extension of weapons is in intercontinental nuclear missiles. These represent technical capacities far beyond even the tools and weapons at the beginning of World War II, and the jumps are surely orders of magnitude greater than the jumps to the stone tools of early human beings, 2 or 3 million years ago. Their importance for the million-year future can scarcely be exaggerated.

Row 6, Detection and Signaling. Probably the greatest early jump in this area came 600 million years ago with the evolution of vision from primitive eyespots to image-forming eyes that could see objects at a distance. Predators and prey could lie in wait, anticipate and plan, and develop a time-organizing nervous system. Sight had to come before foresight, so to speak. For human beings, the great further step was speech with symbolic language, and then writing and printing for distribution.

Today we have radar (which can see through night and fog), powerful lasers, and television. As Teilhard de Chardin said, "The discovery of the electromagnetic spectrum was a tremendous biological event." Television extends our vision, absorbs our interest and emotions, and makes us one humanity around the globe, with simultaneous sharing. We all walk on the moon together or watch national leaders embracing in Jerusalem. Most viewers watch television for four hours or more per day —half of their leisure time—and the Olympic Games in Moscow in 1980 may be watched by some 2 billion people, nearly half the human race.

Television is still spreading rapidly because it is the cheapest way of spending time that the human race has ever devised. Basically it costs only a few cents per person per day—a hundred times cheaper than a car and a thousand times cheaper than a teacher. It is not linear and analytical like speech and print, which are handled by the left side of the brain. With its high-information, holistic, moving-field patterns in space, it is the first invention that has amplified the powers of the right brain, as Marshall McLuhan has noted. In its prospects for individualized storage and recall, and linkages with computers for electronic games and education, its implications for the future are immense. We may come to view it eventually as a greater evolutionary jump than speech and even vision, in its effects on our senses and the way children develop and use their minds, with simultaneous seeing and sharing everywhere.

Row 7, Problem Solving and Storage. Three major methods of problem solving and storage have been developed in the course of evolution. The first is problem solving by survival of the genetically successful, with storage in the DNA molecular chains. This is a phylogenetic mechanism that may go back nearly 4,000 million years, perhaps even before cells. The second is problem solving by individuals with learning nervous systems, by selection and reinforcement and survival of successful acts of behavior, with storage in neural networks. This is an ontogenetic method, about 600 million years old. The third is problem solving by anticipation, by learning and storing the general laws of science and how the world works, laws that we can apply in advance to predict or control, in specific cases that we have never experienced before. These formal laws began with the Greeks, but have become a flood tide in this century, and their levels of refinement and control are now extended a thousandfold with electronic data processing and feedback control.

The example of Sputnik, the first satellite that was put into orbit (in 1957), shows the difference between these methods. It was not one of ten thousand selected by survival because all the rest failed and crashed; it was not a learning Sputnik, trying first a high orbit and then a low one to find the right path. Both of these methods would have been too wasteful. Instead, it had on board the stored computer programs, monitors, and data processing and feedback controls to steer the rocket and to shut off the motors at the right moment, so that it went into the correct orbit on the first try.

Today, electronic data processing and feedback control are at the heart of science, technology, the military, banking, business, social welfare, government, and all our private transactions. They form a collective social nervous system spreading around the world, for storing and manipulating all our knowledge and our environment. Their importance for human future is at least as great as the development of the first learning nervous systems.

Row 8, Mechanisms of Change. This category is not entirely separate from the one before except in its new social organization. The first mechanism of change was Darwinian variation of genetics and selection by survival. The second was the Skinnerian mechanism of variable individual behavior, with selection by the environmental consequences. In early human beings, the coming of symbolic language and systematic thought, with formal anticipation and planning, has been associated with immense new control over the environment and with a threefold increase in the size of the human brain in the last 3 million years. In the mechanic arts, this flowered into deliberate experiment and invention, and then, in the last century, into systematic research and development teams such as those of Edison and the Wright brothers.

But in the last forty years we have begun to organize large-scale systems analysis and design projects such as the rocket projects, the atomic bomb projects, and the man-in-space projects, in several different countries. They involve hundreds of thousands of coworkers in differentiated tasks for many years, and up to 1 percent of the gross national product of great nations. This is a collective thinking process for creating the future, perhaps as big a jump as the original evolution of thought itself.

AN EVOLUTIONARY SINGULARITY

Are these conclusions reasonable? Some people might say that such comparisons are not possible at all. Others would give good arguments against these particular evaluations—although the relative evaluations could be considerably changed for many jumps without affecting the major thesis here. Also the list of recent developments in the last column of Table 1 is obviously not unique. It has been made somewhat redundant so as to bring out certain comparisons, and it might be reduced. Alternatively, it might be expanded, for example, to include psychological and social developments such as behavior modification or credit cards.

Moreover, all evolution does not take place by sharp jumps, although they have come recently to be more widely recognized. Much evolution may consist of almost continuous small steps, and some of the "jumps" here might have taken tens of millions of years. Furthermore, evolution is not always "upward" in terms of progressive increase in complexity or control over the environment. The top level of the pool of life may behave generally in this way, but many species retrogress, for example to parasitism; and of course some 98 percent of all species have become extinct.

However, some of the jumps, and some of the conclusions from Table 1, surely represent phenomena that would be striking to an observer from another world looking at all our biological evolution from a completely nonhuman point of view. The order-of-magnitude changes are too enormous to be glossed over as anthropocentric relativism.

One aspect that would surely stand out to such an observer is the major environmental impact of recent technological developments that are clearly significant on an evolutionary scale. The accumulation of fossil coal and oil and gas took a 100 million years, and the burning of them in just a few centuries is certainly a singular point in evolution. Today this burning is also expected to double the carbon dioxide of the atmosphere by the year 2000, creating a "greenhouse effect" from solar heating that could warm up the climate by 1° or 2°C. This could melt the polar ice caps and change all our agriculture—the greatest climatic change since the end of the last Ice Age, 15,000 years ago.

In addition, a half million or more species of plants and animals may become extinct—10 to 20 percent of all species—because of deforestation, desertification, and the destruction of their habitats. This will be the greatest change in the forms of life on earth since the extinction of most of the dinosaurs, 65 million years ago. It could have major effects on human food production and the whole ecological balance.

Such a conclusion may be resisted by believers in the constancy theory of history or in cyclical theories or cultural rise-and-fall theories. It is often asserted, for example, that the apparent speeding up and increased scale of technical jumps today is just a "recency phenomenon," largely due to our selective emphasis on more recent events. However, the evidence here has been marshaled in part to show that such a view is no longer tenable. In the light of these jumps by many orders of magnitude in so many areas and this unprecedented coincidence of several such jumps at the same time, we are not passing through some smooth acceleration process like the historical past. I think that anyone who is willing to admit at all that there have been sudden jumps in evolution or in history, such as the invention of agriculture, the Industrial Revolution, or the coming of democracies, must conclude from this evidence that we are passing through another such jump far more concentrated and more intense than these, and of far greater evolutionary importance—and that this is what it looks like from inside.

It is essential to understand this if we are to anticipate, even approximately, the consequences of these events and the kind of future we are moving into. It seems to me clear that the future, involving such intricate networks of new linkages and powers from the individual to the global, will be nothing like the past, and that historical wisdom will be of less use to us than ever before in history.

INEVITABILITY AND SURPRISE

The developments shown in Table 1 have many thought-provoking aspects. One is the combination of inevitability and surprise in almost every evolutionary jump. The inevitability is shown by the many instances of multiple invention or development. Image-forming eyes were developed in four different phyla, wings in four different phyla, and electric organs in many different species. Multiple invention is universal in the history of technology. If a new device is physically feasible and advantageous, there will be many directions from which to approach it.

So Faraday's discovery of the relation between changing electric and magnetic fields led, through Maxwell's equations and Thomson's electrons, straight to radio and television by many paths and through the work of many inventors. If any one of these had been missing, others would have made essentially the same contribution within a short time. Probably nothing less than widespread social breakdown or rigid social denial could have prevented the invention and spread of television.

However, the final result is still surprising, and might not have been predicted by any Faraday or Maxwell except as a possible speculation. Too many terms had to be filled in. The new development has its own new laws and regularities which are usually impossible to predict from its incomplete precursors or substructures at a lower level of organization. As late as 1937, Lord Rutherford, the very discoverer of the nucleus of the atom, said that the idea of freeing energy from atoms was "moonshine" —when Leo Szilard had already taken out a secret patent on a chain reaction, and the achievement of the first atomic pile was only five years away. Furthermore, in the 1960s, hardly a single biologist anticipated recombinant DNA, which was achieved by 1972.

This is the situation described by C. H. Waddington with his concept of an "evolutionary valley" of possibilities, surrounded by a "ridge" or barrier. A species may wander or spread over the imaginary "epigenetic landscape," filling new niches or driven by pressure of other species, until it happens to pass over the ridge line into a new valley; and then it rapidly rushes or evolves "downhill" to make best use of the new invention, so reaching a common final path or river ("chreod," in Waddington's terminology) at the bottom of the valley. So reptiles, birds, and mammals, coming from very different origins, evolved wings of long light bones and cambered surfaces, whether skin or feathers, that look very similar to each other—and to the wings of our own airplanes. They must all fit similar aerodynamic requirements, even though these could not have been anticipated or measured before there were wings. Both inevitability and surprise are in play.

This "equifinality" is not teleological or purposive. There is teleology in the watchmaker, or in any cybernetic system that moves a response or shapes a mechanism toward some previously determined goal. There is also teleological equifinality internally in the development of a flower, or any individual organism which can respond to environmental variations with feedbacks and stabilizations determined by its inherited DNA. However, the equifinality of an evolutionary valley whose downhill paths represent a new evolutionary jump is not determined by any previously encoded information anywhere, since the "valley" is just our conceptual description after the fact.

It is true that we might regard the successful interaction of an organism with a new environmental situation—producing a new locked-in or self-maintaining organism-environment pattern—as being "determined" by the "information" in pre-existing but previously unknown laws of the universe, such as aerodynamics before there was anything that flew. Such a view is philosophically tenable. It would lead to the idea that organisms evolving on this planet or any other must come to a common "knowledge" of this "information," at least in the operational sense of fitting into these laws by survival. On this view, our jumps to nuclear energy and space and recombinant DNA were all eventually inevitable here, or on other worlds, from the beginning of life, and the only question is how many billion years they will take on a given planet. However, such a uniformitarian or deterministic view may be more appropriate to earlier stages of life. It gives little place for new flashes of insight or new efforts at design and organization or destruction which might have consequences for a billion years. Moreover, it is an operationally sterile view, more faith than science, offering us no guidance, since this "information" encoded in higher-order principles, by definition, cannot be known in advance of the jumps.

SCALE-UP AND SPEEDUP

A second remarkable aspect shown by the successive jumps in each row of Table 1 is the steadily increasing scale of the systems in moving to the right-hand columns. A one-celled animal can have no eye. It probably takes a tribe to evolve tools, commerce among many cities to develop ships, and populations of millions to need printing or railroads. It is only nations of tens of millions that can have the resources, technical education, and available scientists to make a successful atomic bomb project or man-in-space project, at least the first time. Smaller groups have little chance of competing. In turn, these new developments produce further increases in the scale of organization, as television and technology networks are linking the world today. National boundaries and small sovereign states are almost certainly incompatible with these new global powers.

The most remarkable aspect of Table 1, however, is the incredible acceleration in all our evolutionary jumps in recent times. The familiar time charts of evolution in texts and museums have dramatized this for many years, and the decreasing durations of the successive epochs in Table 1 reflect this; but the further speedup represented by the forty-year period of the last column of the table is a fairly new realization. Probably the new developments, even if inevitable, would have been spread out over a century or so, but the great research and development projects of World War II hastened their advent and synchronized them.

So 1945, the end of the war, might be called World Year Zero. It was the year of atomic bombs, and to come within a year were the first long-range rockets, electronic computers, and the theory of cybernetics.

Within four years we had mass television, the first oral contraceptives, and the discovery of sexual crossing in bacteria. The new biology was part of the explosion of scientific research growing out of the wartime projects. The last column of Table 1 is actually within a time frame from about 1945 to 1972—World Year 027—though of course important further developments, such as those suggested in brackets, may still lie close ahead.

If the successive jumps of Table 1 are plotted as equal vertical increments on a linear horizontal time scale, the curve rises steeply upward. If a one-meter line represents the 4-billion-year history of life, the human period of about 3 million years is a mark less than a millimeter wide at the end of the scale. Furthermore, if this narrow mark is in turn expanded until it is one meter long, our present epoch of less than forty years is only 1/100 of a millimeter wide at the end of this scale —the width of a knife edge.

We frequently use an S-curve or "limits-to-growth" curve to describe some aspects of our age, such as the leveling off of population growth or power consumption as limits are reached. However, this kind of curve of accelerating development is completely different, as shown in Figure 1.

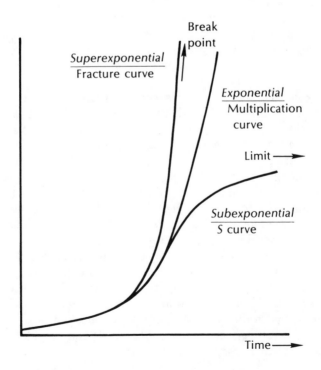

Figure 1. Three types of growth curves.

The limits-to-growth curve is a "subexponential," falling away from the usual exponential curve such as a population curve that doubles in successive time periods. The curve of accelerating developments here is a "superexponential," with its rate of doubling accelerating in each successive time period.

Several phenomena in the natural sciences show superexponential accelerations of this type. One is the fracture of a metal under repeated stresses. The early microcracks grow and merge and increasingly spread with each repeated flexure until there is a sudden complete break of the whole piece of metal. Kenneth Boulding has in fact referred to our present world transformation as a "system-break." Nothing after the break can ever be like what has gone before.

THE MOMENT OF BIRTH

A more illuminating analogy to our situation may be the birth of a baby, as the pressures accelerate in the last days, hours, and minutes before birth. In middle pregnancy, long days pass with one day very much like another. However, when the muscular contractions develop from the maturing hormones and labor pains begin, the mother realizes that this day is different, that the next hour or the next few minutes are not like any hours or minutes before, and that things cannot be stopped, reversed, or even slowed down as they converge toward the moment of emergence.

Such analogies should not be accepted uncritically and should not be pressed too far. Nevertheless, there are some thought-provoking parallels to our present case in this other living and developing system with its own crisis point. We see again the combination of inevitability and surprise. The baby reaches the limits of the womb just as we feel our own pressures of the global limits to growth. Our multiplying problems and tensions may be some sort of birth pangs, as our old theories and institutions prove too small to handle these new realities that have such promise for the long-run future. We may be passing through the birth canal to some new kind of life. In the last ten years, we have seen radical changes in old institutions, such as the family, the church, the school, commerce, government, and diplomacy; and reversals of attitudes toward sex, births, women's rights, business, politics, ecology, work, money, and patriotism, as documented by opinion surveys. Tensions are high, problems multiply, and stable social mechanisms to handle the large global problems are not yet devised. We are suddenly reminded that the moment of birth is the most dangerous moment of the baby's existence. We may ask whether an even sharper crisis of birth is still ahead, or whether we are already through the worst of it and well on the way to a new global society for this new age.

One parallel may be particularly appropriate. In the first few seconds or minutes after birth, the baby must learn to do things it has never done in the womb—to breathe, to cry, to swallow, and so on—and its previous experience does not show it how to do these things. If it fails to do any of them, it dies. So today, we must learn to manage our new powers and problems on a world scale in the next few years—problems of world order, of the environment, of the biological balance—or we may kill ourselves.

The difference is that the baby has information on how to do these new things in its hereditary chromosomes and DNA; it is the descendant of millions of babies that used this hereditary information and survived. The world today has no such inherited information because nothing like this has ever happened before. If we are to survive, we must find out how to solve these global problems by anticipation in advance. We must use all our tools: we must simulate with computers, devise and use the feedbacks for stable collective decisionmaking in technology and society, begin continuous planning, and anticipate problems we have not foreseen. The last two lines of Table 1, the methods of problem solving by anticipation, and global systems analysis, come to be of central importance for our survival.

POSSIBILITIES AHEAD

These evolutionary jumps surely make it more difficult than ever before to predict the social future, even as far as the year 2000. Several years ago, McGeorge Bundy suggested from the increasing rates of crisis that 1989 might represent a point of no return. By that time we might either have collapsed into nuclear escalation, hopeless food and population problems, or economic and political disaster; or we might have established world covenants to manage these problems so that thereafter they would be much less threatening. His date is a magic date, the 200th anniversary of the French Revolution, but if we allow a few years leeway either way, it would fit the timetable of the present analysis.

This means that we may be in a prerevolutionary era in which it is as impossible to predict the year 2000 as it would have been 200 years ago, in 1779, to predict the rise of the French Republic, the Terror, and Napoleon; or, as it would have been seventy years ago, in 1909, to predict the greatest war in history, the fall of the aristocracies, the Communist Revolution, the flapper era, and the Great Depression. Our own future jumps may depend unpredictably on individual acts at critical moments, terrorism, accidents, or the rise of a charismatic leader supported by a mass movement powered by television. It is urgent to set up international checks and balances to help protect us against these instabilities and dangers.

Some social futures might be extrapolated, with much uncertainty, from present trends. These would include such widely discussed changes as individualized television, electronic education, a moneyless society, and world tourism, all perhaps modified by energy limitations. The further transformation of the family, schools, politics, and individual development could be enormous.

The most reliable forecasts, however, are probably to be obtained from just those evolutionary singularities that have made everything else so uncertain. The nuclear dangers, the crisis in oil and coal, and the carbon dioxide problem mean that solar electric development will be steadily favored politically and economically as well as environmentally. Moreover, climatic change and biological extinctions, with their threats to everybody, could be one of the most powerful forces pulling nations into cooperation. If we survive, the international structures of the year 2000 will be very different from their present fragmentary and feeble state.

What may be particularly significant in the evolutionary jumps of Table 1 is the remarkable appropriateness of all the new developments to the prospect of working and living in space.

In space, rockets are the means of travel; data processing and feedback control are the necessary guidance mechanisms; radar, laser, and television are the means of communication for either robot or manned systems; and automation will be the heart of all space factories. Space has no weather and no friction, so that its vast light structures have no stress and can be shaped or moved with essentially frictionless motors. Automatic factories powered by sunlight might operate continuously and unattended for years, for example in integrated chemical processing of asteroidal rocks. Dangerous nuclear or biological experiments might be carried out more safely in space laboratories. The design and maintenance of balanced ecological systems for space construction workers or larger space cities, might be important for understanding and maintaining our ecological balances on spaceship Earth.

However, the controversial proposal of Peter Glaser for satellite solar power stations could be an evolutionary watershed. Solar power is spotty and intermittent on earth, but it is essentially continuous in space, with ten times the power density, and it could be beamed to any country by microwaves, probably with the least pollution of any power source. Many problems would demand resolution, and the first pilot station, utilizing hundreds of construction workers, would necessitate a large capital investment; on the other hand, the project could reveal new technologies and new economies for working and living in space. By the 1990s, this might show whether larger space settlements are technically and economically feasible.

SPACE SETTLEMENTS?

If successful, this technology would be a stepping stone to the bolder proposals of O'Neill. He imagines that there would be rapid multiplication of space settlements, financing their needed supplies from earth by selling solar power—the "gold" of space—and that within a couple of centuries there could be 10 billion people in space and we could leave the earth behind as a wild park, to be visited for recreation and remembrance. The costs per settler and the rates of migration that he envisions are not greatly disproportionate to European migration to America over the last three centuries. The space cities would have continuous and inexhaustible solar power, and the asteroidal rocks could supply a thousand times the land area of the earth. As a result, these cities would be almost self-sufficient, each with its own air, water, agriculture, and recycling. They might be in earth orbits, a few minutes or a few hours apart; or later, they might be in solar orbits, all the way from the orbit of Venus to the orbit of Pluto, and might even be able to adapt to great changes in the sun's radiation.

Just as in the New World, the settlements might explore many different political, economic, and religious systems to give us a great wave of evolutionary diversity and new evolutionary niches. If the earth becomes heavily endangered or contaminated, these could be lifeboats for the survival of the human species.

As the Russian rocket pioneer, Konstantin Tsiolkovsky, wrote in his book *Beyond the Planet Earth* in about 1905, "A planet is a good cradle for the evolution of intelligence; but who wants to stay in the cradle forever?" If anything like such dreams should come to pass, it would mean a cosmic future for the human race far different from anything we have seriously imagined up until now.

But will we go into space? That would appear to depend very much on political and economic conditions and leadership in the next twenty years. If we fail to start these pilot projects within our time, the necessary energy resources, the financial surplus, and the political vision and will may be totally depleted, and the opening up of this frontier may become impossible, perhaps forever. Even if we develop a more stable and hopeful world structure, the needs of the poor on earth, as well as our antitechnological or conservation sentiments, might prevent any consideration of such bold projects for a long time to come. The period from now until about the year 2000 is a "time window" for the initial steps of launching the human race into space, after which it may become impossible.

But the pressures that are imposing "limits to growth" on earth are actually the pressures that will make space ventures more attractive, at least for the next few years. The prospects of abundant clean energy from space, of a virtual lifeboat in case of catastrophe, and of almost unlimited land and resources for the human population without political struggle or war, would seem worth the expenditure of great capital investments, in a time of increasing shortages below. The limits to growth that block us here on earth may actually be a kind of takeoff ramp lifting a technological civilization into space. It is only because of the pressure building up in the seedpod that the pod breaks open and shoots out the seeds. Great technological corporations or competing nations may come to see satellite power stations or space cities as an investment opportunity they cannot resist, and then the race will be on. The next twenty years will decide. We shall then see whether this is a moment of birth into a new form of existence for the million-year or the billion-year future; or whether this is a stillbirth, where the baby dies a-borning because of inability to manage its new problems. A steady-state world on earth may be achievable for the long-run future, where we abandon technological expansion and sustain ourselves with the stimulation derived from aesthetics or inner contemplation; but the abandonment of the larger space venture will still be a great loss for the potentialities of the human spirit.

THE WATERFALL OF HISTORY

Many people hope for great new developments in science, thinking perhaps that our pace of the last forty years will keep up forever; but evolutionary jumps and system-breaks do come to an end. Several of our recent jumps are pressing against the physical or biological limits as we now know them, and there does not seem to be room for other jumps of such a size in the foreseeable future. Other people expect or wish for similar jumps in unconventional areas, such as parapsychology, longevity, or time travel; however, there is not yet much likelihood of such great new powers, and any developments would be too late to have much effect on the next twenty years.

The same pertains to hopes for communication with extraterrestrial intelligence. All of the developments listed in Table 1 seem likely to occur in some form on other worlds in the course of the evolution of life, unless there are unique factors on this earth that we have not yet recognized. This means that life might have reached our current stage of technological development—our knife edge or moment of birth—on millions of solar systems in our galaxy, according to astronomers' estimates of the probabilities of suitable planets. Nevertheless, no signals that would seem to indicate the existence of such civilizations have been detected, although our search for them has not yet been very extensive. Obviously, if any such signals were received, they would completely change our views of the nature of humanity, our future, and our relation to the rest of the universe. But the search for them might have to go on for thousands of years, and in the meantime they can have but little relation to our knife-edge crisis and its resolution here.

This all means that there may not be many great new surprises that would affect our view of the next few decades. We begin now to see the full scope of our problems and possibilities; what tools we have for dealing with them are already in hand. To change our simile once more, we are in the midst of a great rocky rapids or waterfall of history, and we have been rushing toward it almost inexorably for a long time. There is inevitability and surprise, but it is obviously urgent and essential to see how much of this newly envisaged future is really inevitable or really surprising, and how much is still subject to human social design and control. As when going through big rapids, the rule must be, "Don't push the river, but steer the canoe." We cannot turn back, but we can try to use our full new knowledge and powers to steer away from the clear dangers and into more constructive directions. We may yet find ways to cooperate in the solution of the hard collective problems that these new powers have thrust upon us.

If we can solve our problems of global survival in the next twenty years, we may still succeed in breaking through to a new world of abundance, of communications, and of life in space, thus attaining a new social form of human existence for the million-year or the billion-year future as we reach out to the stars.

Chapter 14

Discussion of "Eight Major Evolutionary Jumps Today"

Summarized by Andrei S. Markovits and Karl W. Deutsch

There was some disagreement and dissatisfaction with the optimistic, technologically oriented presentation by John Platt. The discussion was launched by Joseph Weizenbaum, who commented that the analogy Platt had drawn while presenting his paper, between the incredible changes of our modern era and a thunder clap, reminded him of Sam Goldwyn's reputed response to the news of the atomic bomb. "The atomic bomb?" Goldwyn asked. "That's dynamite!"

Harrison White came to the defense of economists, who he thought must have been made "quite uneasy" by Platt's presentation. "Markets and industrial systems of various kinds, which are social inventions of some considerable importance, and upon which many of the developments you outlined depend," he told Platt, "have been slighted."

Platt replied that, while he has written about these topics in the past, they were not à propos to the topic of this conference. Many members of the panel disagreed, however. Platt continued to defend his paper by insisting that no present social developments were on an order of magnitude beyond those mentioned in his paper.

White retorted by saying that he felt there was a "real disagreement" between himself and Platt. "I am appalled by the scale of values you [Platt] bring to these things," he said. White felt that Platt lacked a sense of scale or true priority. Though he acknowledged Platt's leadership in trying to have the group think more boldly and on a larger scale, White nonetheless could not "resist the idea that there is here a living embodiment of what some previous speakers and commentators have referred to: a kind of rather over-focused technological élan, a sense of a technological fix, which seems to me to be rather worrisome." White supported his statement by suggesting that, for instance, the Pyramids were of equal importance in technological achievement to Platt's example of Project Apollo.

Karl Deutsch discounted White's example because, in Deutsch's view, the Pyramids did not constitute a new environment in any way. They did not conquer any new space and thus could not be compared to our current space projects. White, however, defended his example, suggesting that, to the people of the time, the Pyramids opened up a new dimension of perception, thereby increasing the Egyptians' sense of space by "an uncountable order of infinity."

Gerhard Leminsky expressed his difficulty with Platt's paper by attempting to complement Platt's way of thinking with a political dimension which he [Leminsky] found lacking in Platt's scheme. At the end of his paper, Platt said we must do something. But this seemed to Leminsky to be an appendix, an afterthought without substance. Platt spoke as if things were proceeding in a natural way; there were neither social conflicts nor political choices in his framework. All seemed to evolve naturally, logically, and compellingly. This, Leminsky argued, was a very static way of looking at the world's problems—curiously ahistorical, despite Platt's impressive excursions into the earliest periods of our existence.

Leminsky perceived a very sharp contrast between the statements made by Platt and by Weizenbaum. He felt that it was not important that the former's view is optimistic and the latter's pessimistic; what is significant is how the possibility of political intervention might be introduced into Platt's system of thinking. Leminsky suggested that to accomplish this, Platt would need to include a "systematic embodiment of social invention" in his scheme.

Platt replied that he was describing two things: transitions and, more importantly, the uncertainties associated with any new form of life, which also includes the beginnings of social structures. All beginnings are a moment of birth. Birth is, indeed, the most dangerous moment in a baby's life. "If we can get through this birth canal in which we presently find ourselves," Platt said, "and actually design mechanisms for solving the problems of our collective systems in the next twenty years, then we may have a long future ahead." Platt cautioned, however, that the design of those mechanisms is of most urgent priority.

I. Bernard Cohen pursued Leminsky's point by agreeing with him wholeheartedly that Platt's paper was marred by a lack of historical perspective. There were two problems raised in Cohen's mind. First, he held that Platt left out "the whole historical process between the time that man, so to speak, came out of the tree, and the present." Cohen maintained that there were, during these years, "not only simple changes, but social changes which also have to be quantified." Furthermore, Platt's scheme of quantification failed to take into account crucial developments, at least of a similar order of magnitude to those described in his paper. Cohen cited the invention of language as a very important event. He believed that Platt ignored tremendously significant social developments in the history of man, such as the developments of agriculture, metallurgy, and urbanization. He criticized Platt for starting at a biological beginning, proceeding to a technological present, and leaving out everything in between. Cohen concluded, "I must say, I find it quite shocking. It is almost as if you would take the position, and I am sure you do not, that after biology, all of man is embodied in present technology, and that a solution to all his problems must be based only in these terms." Cohen judged social development, along with all human development, to be absolutely critical.

Leminsky concurred with Cohen, reiterating and elaborating upon his earlier criticism of the absence of political considerations in Platt's paper. There was no room for social and political intervention in the way Platt presented the existing structural relations and future developments, Leminsky agreed. With power relations unmentioned, it was unclear who controlled whom, how, and for what purpose. Leminsky asked again, how then do we effect change?

Claus Offe continued along the same lines of criticism as Leminsky's. He felt that in Platt's scheme, political power was missing and that, furthermore, this omission led to an equation which, in Offe's view, is "fundamentally false." Platt's evolutionary stages are not one continuous process. Offe elaborated:

> They differ in that the first [biological] is not mediated or propelled through any decision, whereas the second [technological] is. What do I mean by "decision"? Decision is an outcome, an event that could be otherwise. It is contingent in the sense that it allows for difference. There is a degree of freedom involved for the second group, whereas the first group of evolutionary processes is not free.

In equating and collapsing these two forms of evolution—the biological and the technological—Platt discounted human choice and social action in the shaping of history and destiny, Offe concluded.

Karl Deutsch discussed the idea of a watershed. (Mathematicians call this concept a separatrix.) If this process is in the realm of time, we call it a crisis. To illustrate his point, Deutsch used the example of an airplane going down a runway for takeoff. It goes faster and faster, approaching a critical, qualitatively different point: the "watershed." At the moment at which it is airborne, it is controlled by the laws of aerodynamics, which are quite different from those that control it on the ground. "Nevertheless, for certain purposes, some things that happened before the plane lifted into the air and after it was airborne, are comparable. For other purposes, they are not."

Deutsch felt that it was ridiculous to adopt a "thou shalt" or a "thou shalt not" attitude, because for certain things a comparison will be valid, and for others it will not. Therefore, one should not dictate beforehand that *all* comparisons are valid or that *all* are invalid. Deutsch went on to note, "Platt argues that never before have so many things changed within so short a time. Furthermore, Platt submits that this is an unprecedented thing and, even more importantly, that we are right in the middle of it." Continuing his analogy of the airplane during takeoff, Deutsch said that while we are arguing about whether the plane will ever take off or not, the fact is that the plane is already airborne, and the task at hand is to see that it does not crash.

Deutsch held that a more relevant question for discussion was whether or not we are already far enough off the ground so that with reasonable certainty we can say that we will not crash. "The question of whether you are airborne or not," he said, "is analytically distinct from the question of whether you will crash or not. Platt argues that we are right now in the middle of passing through one of the huge watersheds of the evolution of human history, and that much of the present worry and anxiety people are experiencing is very appropriate and understandable, because that is exactly what is happening to us." Deutsch suggested that the key argument should be over whether we are in the middle of this or not. "Is our present time an unusual concatenation of major changes defined by qualitative jumps best described by Hegel's view of history?"

Peter Weingart then commented, "It is not often that one learns something at a conference, but throughout these meetings, I thought and spoke along the same lines as Offe and Leminsky, and your [Deutsch's] argument against those views, I must say, I have to accept." He went on to formulate a contradiction which he felt was present in Platt's scheme. So far, the views of catastrophists and optimists had been discussed, with the former saying that it does not matter what we do, it will all end badly anyway, and the latter maintaining that we should not worry, everything will turn out all right. Weingart turned to Platt's Table 1 and looked at the comparison of the development of thought and the development of electronic data processing and feedback control. "Is the development of organization on a social scale equivalent to the development of thought?" he wondered. Thought has kept mankind out of many troubles. If feedback mechanisms are the functional equivalents to thought, they should contain a third possibility between "mindless optimism" and "mindless catastrophism." It should enable man, on a social scale, to develop mechanisms which promote a development of science and technology that will certainly not be self-destructive. "If it were self-destructive, we would be on a scale with dinosaurs. We would simply die out because of hypertrophic abilities that will develop and kill us all." Mechanisms can be developed, however, that will keep man from destroying himself. Weingart conceded that this is still deterministic, but it would enable one to live with Platt's ideas more easily and introduce such things as free will and the ability to cope with one's own fate.

Platt was given the chance to defend himself against his various critics. He began by commenting that he hoped some were right in claiming that we may not be in *the* "singular epoch" and that we may have more time than he supposed. He was distressed to hear that historians do not know whether or not societies are in an evolutionary jump until after the facts, in a completely *ex post facto* fashion. This remains of little use to policymakers who are trying to improve the present and the future in tangible ways. He said he hoped that historians would sharpen their perceptions so that their insights would yield concrete benefits for the decisionmaking process.

Platt addressed Cohen's comments about the absence of the social dimension. He felt that he had adequately included the important aspects of human evolution in the last 3 million years, in three columns of Table 1. All those aspects that various critics singled out were mentioned in the table: language, agriculture, metals, urbanization, the scientific revolution, the industrial revolution, and the development of invention. Platt agreed that they all played an important role in the creation of the present human condition.

Platt moved on to White's comments. "White is right to state that Project Apollo was certainly not the first great organization of human achievement. It probably represents the highest level of organization in terms of technology, however. But in terms of a unique qualitative jump compared with that which went on before, it may not be such a great jump, as I tried to convey in my paper." He also agreed that, for their era, the Pyramids were certainly a great technological and organizational achievement.

Platt, however, took issue with White's comment that there were no ethics in his scheme. Others had commented that there were no political dimensions, human effort, values, or choice in his system. They seemed to see it as a technocratic automatism deterministically describing the evolutionary process. Platt defended himself against these accusations, citing in particular the section in his paper entitled "Inevitability and Surprise." One of the reasons why he wrote the paper, he explained, was to distinguish between those aspects of the present situation—the watershed aspects—that cannot be changed, and those that can. It is important first to be aware that there are some aspects in our present situation which are almost inevitably irreversible, and second, to separate them from those that can be affected by political choice. He continued, "The ethics that guide me are essentially the ethics of survival."

Platt noted the presidential address to the American Psychological Association given by Donald Campbell. It dealt with the sociobiology of ethical and religious principles. Platt believed that it raised questions about the origins of our ethics. Essentially, our ethics are those of the "tribes" that survived. When a baby is about to be born, our first act of ethics is to ascertain its health in structural and systemic terms. Then our larger ethics, the development of experience, can come later, if the little creature survives. "So today," Platt continued, "we have to think in terms of a global system constructing its feedbacks, checks and balances, and self-stabilizing mechanisms, thus inventing and implementing them for survival. That constitutes the highest ethics for the next twenty years."

Platt then addressed the question asked by many of his critics, of how we can switch from a deterministic and automatic historical evolution—which has no aspects of human choice in it—to a condition in which choice in general and choice in policy formulation and implementation become major determinants of the future. He compared the question to a situation in biology today.

Human things have become the agents of evolution. In the course of evolution, to go back to the religious metaphor, man is created in the image of God. Some have turned it around and said that God is created in the image of man. That's all right, too, because we could not have these concepts of a God who planned and who shaped the long-run future unless human beings had had such a concept themselves. If man is created in the image of God, then God works through human beings and our hands are His hands, in a sense. As we develop into the conscious agents of evolution at the biological level, we also simultaneously become the same in the political realm. It is a synthesis of the two that will see to it that these present birth pains we are experiencing are channeled into the proper direction so that the human race, its goals, and its experience survive in the long-term future.

Deutsch then proposed that an addendum might be attached to Platt's table. He suggested a ninth function in addition to the existing eight, namely that of "social implementation." He explained how this might fit into Platt's scheme. "We have not made major progress in obtaining consensus, cooperation, or coordination on a large enough scale and with enough reliability to deal with many of the problems that come up." Deutsch then gave his summary of Platt's message and that of the conference as a whole: "It is not, 'Don't worry, everything will be all right,' nor is it 'The world is about to end,' but, rather, 'Tighten your seatbelts, extinguish your cigarettes, get ready for takeoff, and be prepared for some expected turbulence ahead!"

With this, the discussion came to a close.

PART III

Critical Discussions for the Future: What to Do? Who Should Do It?

Chapter 15

Fear of Science Versus Trust in Science: Future Trends

Max Kaase

INTRODUCTORY REMARKS

Processes of educational upgrading, extension of mass communication networks, and increasing politicization of mass publics—as some core concomitants of general sociopolitical change in advanced industrial societies—all contribute to a situation in which public anticipation of particular developments can become extremely important for political decisionmaking. Science and technology are two cases in point, as the current debate over nuclear energy indicates. Lack of the kind of systematic empirical evidence about the past, which social science (through survey research) can now produce about the present, makes it difficult to assess previous distributions of fear and trust in science, as well as the conditions for their prevalence—if ever, in human history up to the present, such a set of crystallized attitudes toward science existed at all. Let us assume, however, that such attitudes exist at least now and describe their structure. We may then proceed to speculations about the future trends of such attitudes and the conditions that may shape these trends.

PUBLIC ATTITUDES TOWARD SCIENCE

The following analysis is much hampered by two constraints that often interfere with an adequate assessment of social phenomena: lack of theory, and lack of data. Thus, precious little theorizing about the factors that shape mass attitudes towards science is available, and empirical data of any consequence for the topic are also scarce.

It may well be argued that fear of science pertains to a very special aspect of social reality and is therefore not related to any general feeling of fear or trust individuals may experience in the present. We have no way to answer this question empirically. There is, however, sufficient evidence to corroborate the point that contemporary mankind, even in the industrially developed countries, has not ceased to experience fear of the future (Noelle-Neumann, 1976:294). The relativity of expectational standards, archaic fears of evil forces, but also—through the media—a broadened but at the same time increasingly indirect view of the world, selectively shaped by the negative attention standards of the news-makers, may be factors in this sustained fear. The electronic media in particular have helped to produce an enormous public response to such spectacular events as nuclear disasters, terrorist attacks, or airplane crashes. Such events relate directly to surges in the level of fear, even if these surges are only short-lived.

With regard to specific attitudes towards science, we can fortunately rely on some data collected from two surveys carried out in the spring of 1977 and in October of 1978 in the nine countries of the European Community and commissioned by the EC Council (Kommission, October 1977; Kommission, February 1979). One of the many problems with these studies is that they use the concept of "science" indiscriminately from the various institutional contexts in which "science" can be embedded—be it independent research at universities or research institutes, research at private enterprises, or even government-controlled military research. Thus, the overall positive assessment of past and future effects of science on the lives of people that emerges from these studies is certainly related to the fact that the concept of science is for many people still inextricably intertwined with the institution of a free university of the Western type.

A first question in the EC survey dealt with the general attitude of the citizens towards science. Unfortunately, the question itself is quite problematic because of dubious dimensionality underlying the closed-ended response continuum:

Science

   only satisfies the curiosity of scientists
   is a major factor in the improvement of the human condition

carries more disadvantages than advantages
is dangerous
is excitingly interesting
don't know/no answer.

Despite this unsatisfactory mix of response categories, a little less than 80 percent of those EC citizens answering the question agree that science is a major factor in the improvement of people's lives.

A second set of questions deals with the public's perception of whether its life has experienced major changes during the last twenty-five years, and of whether these changes were to the better or to the worse. Whereas there is almost unanimous agreement that such changes have occurred, about one-half of the EC population feels that those changes were to the better, one-fourth thinks they were to the worse, and the rest cannot make up its mind. What is relevant in the context of this discussion is that, among those who have perceived changes over the last twenty-five years, nearly all agree that science has been an important factor in those changes.

The general positive assessment of science is further substantiated when we consider the responses to the question of whether in the future there will still be good and useful things to be discovered by science. Here again, excluding those without opinion, there is almost unanimous agreement that science can indeed continue to make useful discoveries. Finally, again almost nine-tenths of the EC population are in agreement that science can also contribute to the betterment of living conditions in the developing countries (Kommission, 1977: 36 ff.).

Without a detailed secondary analysis of the data, it is impossible to determine whether respondents with negative attitudes towards science show a particular, clearly identified profile. With data so close to unanimity it is, however, very unlikely that we can detect such profiles for statistical reasons alone. Thus, it seems fair to interpret these results as reflecting a general consensus on the importance and benevolence of science.

It would be naive, though, to assume that a mass public that has become increasingly attentive to and skeptical of the way in which politics deals with scientific and technological discoveries (such as nuclear fission and nuclear plants) would rest its case at this point. To the contrary, there is full agreement among the nine European nations that, even without consideration of potential military applications, scientific discoveries may have dangerous effects. In addition, there is a clear understanding that developments in science and technology involve increasing risks and are difficult to control. As the EC report summarizes: "The results of this test are clear: they reveal a broadly generalized climate of fear." (Kommission, 1978: 42; my translation, M.K.).

Table 1. Attitudes Toward Science in the Nine Countries of the European Community

	Belgium (%)	Denmark (%)	Germany (%)	France (%)	Ireland (%)	Italy (%)	Luxembourg (%)	Netherlands (%)	United Kingdom (%)
1. General attitude toward science									
Science is a major factor in the improvement of the human condition	81	75	76	75	65	83	83	78	72
Percentage of missing data	12	11	22	5	6	6	3	6	7
2. Perception of life changes during last 25 years and impact of science on those changes									
Yes, life has changed	96	97	86	96	97	96	97	95	96
If life has changed:									
Changes to the better	56	59	67	49	75	48	71	45	50
Changes to the worse	21	16	15	24	17	24	10	22	33
Undecided	23	25	18	27	8	28	19	33	17
If life has changed:									
Science has had major impact	93	86	93	95	85	89	97	95	87
Percentage of missing data	6	7	3	2	8	5	6	2	5

3. Potential of science for future advantageous discoveries

Yes, potential is there	97	93	95	98	98	98	96	97	98
Percentage of missing data	11	14	13	7	9	5	5	8	6

4. Potential of science to improve living conditions in countries of third world

Yes, potential is there	89	84	79	89	88	90	88	85	87
Percentage of missing data	10	14	19	10	11	9	11	14	12

Source: Kommission, 1977: 36–40.
Note: Percentages have been recalculated excluding missing data. Missing data percentages, except in cases where few missing data occur, are given in order to assess the magnitude of this factor.

Table 2. Attitudes in the European Community Regarding Potentially Dangerous Effects of Science

	Belgium (%)	Denmark (%)	Germany (%)	France (%)	Ireland (%)	Italy (%)	Luxembourg (%)	Netherlands (%)	United Kingdom (%)
1. Existence of dangers[a]									
Dangers exist	89	82	83	75	70	66	86	81	81
Dangers do not exist	11	18	17	25	30	34	14	19	19
Percentage of missing data	18	15	20	12	18	17	7	11	11
2. Risks and difficulties of science and technology increase[b]									
Agree	81	85	84	85	80	85	84	77	76
Disagree	19	15	16	15	20	15	16	23	24
Percentage of missing data	26	29	25	15	26	15	24	14	15

[a]Source: Kommission, 1977: 37.
[b]Source: Kommission, 1979: 31.

Note: Percentages have been recalculated excluding missing data. Missing data percentages are given in order to assess the magnitude of this factor.

182

At this point it may be useful to consider a conceptual differentiation between science as the institutionalized production of systematic knowledge, and technology as the application of scientific knowledge to achieve practical ends. Without using this terminology, people, as the data from the 1978 EC study clearly show, direct a substantial skepticism towards the way scientific discoveries are *used* (Kommission, 1979: 31). This interpretation is further substantiated by data coming from the 1977 EC study. When asked where—under the condition of scarce resources— funds for scientific research should be directed, four major areas, all related to the natural, medical and biological sciences, emerged:

medical research
research to improve worldwide supply of food
control and reduction of environmental pollution
development of new forms of energy (Kommission, 1977: 53 ff.).

Surely, science is overwhelmingly associated with and evaluated through its impact on the improvement of the quality of life today. That it may already have gone too far has, in modern times, been a constant topic for discussion and reflection after the Hiroshima bomb; actions of civic initiative groups, the Three Mile Island incident, and related reports of malfunctioning nuclear plants have certainly helped to create apprehension about this aspect of science among mass publics (Lenk and Ropohl, in: Hammerich and Klein (eds.), 1978: 282).

This apprehension is strongly reflected by the expression of fear by four-fifths of the EC citizens over the ongoing destruction of nature by pollution. But other technological developments as well are of concern to the people: automation and its effects on unemployment, increasing artificiality of living conditions, and the impact of medical and pharmaceutic discoveries on people's health (Kommission, 1978: 32 ff.)

It has been frequently argued before that mass publics are unable to assess complicated matters competently. Obviously, this statement would then be particularly true in reference to something as complex as science and technology. However, there exist good reasons and plenty of data to prove that the public may possess better judgment than those critics are willing to concede. Surely, in-depth probing would reveal that, on the average, citizens do not possess detailed, contextual knowledge about those subjects. But, as the following results of the 1978 EC study indicate, the European mass publics have a clear understanding of which research projects entail unacceptable risks and which may pay off. Eight research projects presented to the respondents in closed-ended list format were regarded as potentially dangerous in the following rank order (Kommission, 1979: 65):

1. new sources of energy   5%
2. organ transplants   7%
3. synthetic raw materials   12%
4. research satellites   13%
5. genetic experiments   35%
6. nuclear power   36%
7. centralized data holdings   45%
   on citizens
8. synthetic food   49%

In every instance, it is the interference in the natural and technological environment of the citizen that is regarded as potentially threatening.

The data presented so far seem to permit the conclusion that the very favorable picture citizens of the nine EC countries held of science in 1977 and 1978 reflects a latent distinction between science and technology. In addition to the separation from application contexts, the high status of the scientists as scholars as well as idealized concepts of the hard-working researcher may have helped to establish this positive image. Should, however, there occur spectacular events—in particular, catastrophies—which could be attributed to science and research, then an entirely different public perception of science might rapidly emerge, with potentially lasting consequences for the *institutional makeup of science*, because of the political relevance of such events and the need to cope with them politically. This is a concern to which we will return shortly.

At this point one qualification of the preceding analysis seems in order. All of the countries in the EC are industrialized countries—with varying levels of affluence, of course—and fall into the category of democratic regimes. Thus, variations in economic development and political order cannot be studied in terms of their effect on attitudes towards science. Fortunately, an empirical study (of slightly doubtful methodological quality) dealing with the public's image of the world in the year 2000 conducted in eleven countries from 1967 to 1969 may help to qualify and differentiate some of the findings of the EC studies (Ornauer et al., 1976). The "Year 2000" study itself contains only two sets of questions with regard to the impact of science at all: the potential for and the desirability of certain scientific discoveries or developments. (For the marginals of those questions, see Ornauer et al., 1976: 656 ff.) Four findings from that study are relevant to our discussion.

First, there is a substantial negative correlation between skepticism towards science—defined as a negative ratio between *likely* and *desirable* outcomes of scientific activity—and economic development: the more highly developed the country, the more skeptical its outlook on science

(Galtung, in: Ornauer et al., 1976: 57 ff.). In other words, as nations develop economically and change socially, their population becomes increasingly aware of potential negative consequences of science. An analysis of student samples in four countries accentuates this picture even more greatly: "The group from which the future scientists will be recruited is disillusioned with science, seeing it either as dangerous or as futile." (Wiberg, in: Ornauer et al., 1976: 195)

The second relevant result of the Ornauer et al. analysis is that differences in the political order do not supersede those differences just described. Thirdly, and especially relevant for the social sciences, a particular skepticism towards science seems to exist with regard to the likelihood of gaining social achievements, such as the cessation of wars (Galtung et al., in: Ornauer et al., 1976: 567). Finally, the authors identify one critical point for the future assessment of science: the danger of an increasing penetration into the *"inner sphere of human existence,"* (Galtung, in: Ornauer et al., 1976: 63; his emphasis), operationalized by the determination in advance of the sex of children, and the preselection of their major personality traits.

Summarizing the first part of the analysis, we can conclude that, probably varying by level of economic development, there still exists a strong tendency to evaluate favorably the institution of contemporary science. This evaluation reflects the public's sense that science has been and will continue to be a major factor in improving people's living conditions.

However, there is a definite understanding among mass publics that technological derivatives of scientific discoveries—but also certain scientific discoveries themselves—may entail dangerous consequences. A case in point is the field of nuclear energy, where innocent optimism has declined and skepticism now prevails. Even in this instance, however, the view of the Europeans is quite complex and balanced.

First, while *on the average* there is a slight majority of people who see positive payoffs in the further construction of nuclear plants as compared with those who see unacceptable dangers, this is not an *overall* majority (44 percent). Secondly, opinions in the nine countries are heavily split, with strong skepticism in Belgium, Germany, and particularly the Netherlands, and quite a bit of optimism in Italy and the United Kingdom. Thirdly, there is full recognition of the risk one is taking by delaying the construction of additional nuclear plants, a risk the consequences of which are regarded as quite serious by the majority of those Europeans who express an opinion on the matter. Thus, it may be safely stated that, even with a substantial skepticism towards technological developments and—held by more than half of the Europeans with

Table 3. Attitudes Toward Nuclear Energy in the Nine Countries of the European Community

	Belgium (%)	Denmark (%)	Germany (%)	France (%)	Ireland (%)	Italy (%)	Luxembourg (%)	Netherlands (%)	United Kingdom (%)
1. *Attitudes toward further development of nuclear plants*									
It pays	35	46	40	44	50	59	40	30	62
It carries unacceptable dangers	48	43	52	47	39	32	35	57	27
Does not pay (is un-interesting)	17	11	8	9	11	9	25	13	11
Percentage of missing data	18	20	13	10	10	10	12	6	8
2. *Assessment of consequences of construction halt for nuclear plants*									
Will force consumers to re-duce energy consumption	59	59	57	63	64	73	53	71	69
Will not force consumers to re-duce energy consumption	41	41	43	37	36	27	47	29	31
Percentage of missing data	22	12	21	13	16	11	13	8	12
3. *Risk assessment of construc-tion stop for nuclear plants*									
Very risky	47	54	63	57	66	65	41	49	70

Source: Kommission, 1979: 69

Note: Percentages have been recalculated excluding missing data (except section 3 where data for recalculation were not available). Missing data percentages are given in order to assess the magnitude of this factor.

opinions (Kommission, 1979: 32)—a romantic desire "to return to nature," there is a strong sense that *controlled* technological achievements are mandatory.

From analyses of the process of public opinion formation, it is well known that opinions can fluctuate substantially depending on contextual circumstances and, in particular, as a result of events. Since attitudes toward science and technology are presumably not very central to individuals' belief systems, it is plausible to expect that those attitudes will be particularly susceptible to external factors. To what extent this flexibility constitutes a potential threat to institutionalized science is difficult to assess. Obviously, this threat could only be mediated through the political system. For those who value highly an independent science, it will be good news that, according to the EC study, attitudes towards science up to this point—with the slight exception of nuclear energy—are not yet politicized (Kommission, 1978: 64). On the other hand, the ability of mass publics increasingly to apply direct action techniques to influence political outcomes (Barnes, Kaase, et al., 1979) may, for instance in the cases of a nuclear catastrophe, biological poisoning, or adverse genetic experiments, demand and quickly obtain major changes in the institutional makeup of science. Such a process may well be supported by the political elites who, for various reasons, show an increasing inclination to interfere in institutionalized science.

CHALLENGES TO CONTEMPORARY SCIENCE

In West Germany, the freedom of science and research is guaranteed in Article 5.3 of the Constitution, thereby reflecting the importance given to a free science in a free democratic society. Interestingly enough, a discussion triggered by a seemingly nonrelated, highly controversial issue—the 1978 German data protection law (*Bundesdatenschutzgesetz*)—has made many scientists suddenly aware of the fact that freedom of research and science is in competition with other basic rights, in particular the right of the individual to have his sphere of privacy protected against intrusion from the outside (Simitis, in Kaase et al., eds., 1980).

Of course, it is not just a coincidence that this discussion followed a specific path, namely the path—especially with regard to the social sciences—of increasing government intervention, not so much concerning the dissemination of data as the tendency to censure research topics. Why is that so? In this article it is claimed that some of the forces behind this particular development bear a more general significance for re-flections on the future status of independent science and research.

The survey evidence presented in the previous section of the article seems to indicate that, while there may be a certain (increasing?) skepticism toward science and even more toward technology, science still enjoys a high legitimacy and status. By contrast, as Scheuch (1979: 3) argues, in the eyes of the governing elites in Western industrialized countries, science *as an institution* is experiencing an increasing de-legitimization or, in reference to Tenbruck, even a trivialization. Why should that be?

One first factor is that the industrialized countries of the West are experiencing, as Inglehart (1977) argues, an ongoing process of value change, from materialist to postmaterialist values. One concomitant of this process is a rising pluralization and democratization of values which exposes established institutions like science to influences from the outside world, for better or for worse. This phenomenon is clearly exemplified by the efforts to retailor the decisionmaking processes in German universities to resemble more closely those in the political field. Another aspect of this development is to impose from the outside—and accept from the inside—the principle of "social relevance" as a yardstick for good science and research.

That changing values create problems for the traditional role defini-tions of scientists becomes obvious when—as brought home to scientists' minds in Germany by the data protection discussion—one examines the kind of experiments scientists conducted without any regard for poten-tial consequences for their subjects: the famous syphilis experiment with blacks in the United States in the thirties, the Milgram experiments, and so on. Surely, these may, even then, have been extreme cases, but contemporary evidence hints to the fact that, at least in Germany, the various academic professions have not sufficiently concentrated on the protection of the rights of those they experiment with, in true experi-ments or even in the less problematic case of field experiments. One example from the social sciences may further help to illustrate this point. In democratic societies, free cooperation of those participating as test subjects in research has been an essential prerequisite—the so-called "informed consent." Frequently, the "information" given to the partici-pants has been incomplete, superficial, even outright misleading. Where-as in the United States "codes of ethics" have been established as mechanisms of professional self-regulation, a similar discussion in Ger-many started only recently, triggered mostly by the impending new data protection law. Thus, it is not surprising at all that Spiros Simitis, the data protection officer from Hesse, has called "codes of ethics" the last resort of professions which are losing their advantaged status (Simitis, in Kaase et al., eds., 1980).

While these examples point to scientists who, obsessed by their quest for knowledge and unimpressed by the possible consequences of their scientific vigor, have experienced and are continuing to experience a decline in the legitimacy of their profession, the challenge to science as an institution is by no means exclusively caused by malfunctions within science. One other major factor well worth considering is that, because of the increasing scope and consequences of scientific research, its results become more quickly and more directly relevant to the distribution of power in societies. Not only can science help or be believed to help to increase and preserve power (particularly political power), it also can help to destroy power. Thus, it is only consequential that increasing political influence is used to steer and control scientific research.

This control is considerably eased by the fact that most of the money made available for scientific research is no longer channelled into science through processes of independent peer review, but rather through direct government spending or through private enterprises, both operating outside the system of peer review and control. As a logical consequence, more and more research is conducted under strict legal and procedural regulation.

This development, however, may only in part be interpreted as the result of a lack of professional self-control on the one side and power-hungry politicians or civil servants on the other. Rather, the potential scope of scientific discoveries has by now become such that it seems pertinent to establish criteria and institutions which, even if they cannot stop the process of discovery, can at least successfully control the use made of those discoveries and prevent their undesirable and often unanticipated consequences.

One final aspect briefly mentioned before refers to the fact that, because of socioeconomic and technological developments, the direct impact of science is no longer felt only by the elites, but also by the mass publics; it may thus trigger broadly based reactions. In particular, the spread of direct action political techniques has created the possibility for ad hoc action elites which, with assistance by the media, can react swiftly and competently outside the slow process of parliamentary democracy.

Outlined very sketchily, these are some of the dilemmas contemporary science is facing. Hardly noticed by the public, the institution of science has not only gradually changed its structure, but also has experienced changes in the guiding principles under which it operates. Science is now a controversial enterprise, ridden by value conflicts, shaken in its belief in the validity of the central principle—search for truth—and delegitimized through the destruction of the ivory tower. What shall be done?

THE FUTURE OF SCIENCE

It is a matter of speculation to what extent Western democratic societies will be able to continue as free societies. Similarly, for reasons previously discussed, it is unlikely that science will be able to continue as an institution guaranteeing individuals the resources, and the isolation from external interference, necessary to foster scientific discoveries. The neo-Marxist perspective of "social relevance" (*gesellschaftliche Relevanz*) as a yardstick for the evaluation of science has led many observers to conclude that external control of institutionalized science is the proper device to ascertain that scientific research has a maximum payoff for society. The history of science as well as the theory of science indicate that this perspective suffers from the fundamental illusion that scientific progress depends on the formulation of practical problems (Albert, 1979). However, the broadening potential consequences of scientific research will undoubtedly make it necessary to set up institutions of professional self-control, which assess research in light of its consequences for society, groups, or individuals, and which learn to perform responsibly in situations of value conflict.

Science as an institutionalized search for truth can only successfully survive in a nontotalitarian political order. But even bourgeois democracy is a system of authority relationships (*Herrschaftsform*), and scientists have to realize that science cannot be kept isolated from the sociopolitical processes around it. The effects of science and applied science, we have argued, become increasingly consequential for power elites and mass publics alike, and it would have been naive to assume that, under those circumstances, science could have remained in the ivory tower forever.

What can science and scientists do to ascertain their status as crystallizing agents for the continuing advancement of knowledge? Finally having realized that they do not act in a societal void, for one they will have to build adequate institutions of self-control that will help anticipate and counteract potentially dangerous consequences of scientific discoveries, thereby assisting in establishing new bases of legitimacy. On the other hand, with their old bases of legitimacy deeply embedded in an autocratic, elitist value system rapidly waning under the pressures of participatory democracy, they will have to learn the techniques of collective bargaining and direct political action in order to survive against outside pressures from society. There can be no question that they possess the resources necessary to be successful.

CONCLUSION

From the reflections and data presented in this article, it may have become obvious that the question regarding the trust in and fear of science is too simple, probably even misleading. Science has been and will continue to be generally regarded as a central force behind the improvement of the living conditions of a substantial part of the world's population. However, there is an increasing awareness of negative consequences that follow from and go with the further penetration of rationalism and demystification of the world (*rationale Durchdringung und Entzauberung der Welt*) that renders the basic trust in the benevolence of science conditional on future developments in science and technology. This, in addition to the enormous scope contemporary science is assuming, has moved science into the focal point of socio-political cleavage lines. Therefore, science as an institution and as an individual challenge to those working in it has been removed from the ivory tower. Trust and fear with regard to science are equally justified; both will help shape the future makeup of science. Which of the two will prevail, however, is contingent on many of the factors sketched in this article.

BIBLIOGRAPHY

Hans Albert, "Geplante Wissenschaft: Liberalismus—nach wie vor," in: *Neue Züricher Zeitung*, Zürich, 1979.

Samuel H. Barnes, Max Kaase, et al., *Political Action: Mass Participation in Five Western Democracies.* (Beverly Hills, Calif.: Sage, 1979.)

Ronald Ingelhart, *The Silent Revolution.* (Princeton, N.J.: Princeton University Press, 1977.)

Kommission der Europäischen Gemeinschaften, *Wissenschaft in der Öffentlichen Meinung Europas.* (Brussels, 1977.)

Kommission der Europäischen Gemeinschaften, *Einstellungen der Europäischen Bevölkerung zu wissenschaftlichen und technischen Entwicklungen.* (Brussels, 1979.)

Hans Lenk and Günter Ropohl, "Technik im Alltag." In Kurt Hammerich and Michael Klein (eds.), *Materialien zur Soziologie des Alltags*, Sonderheft 20 der Kölner Zeitschrift für Soziologie und Sozialpsychologie. (Opladen: West-deutscher Verlag, 1978.) pp. 265–298.

Elisabeth Noelle-Neumann (ed.), *Allensbacher Jahrbuch der Demoskopie 1974–1976*, Volume VI. (Vienna–Munich–Zurich: Molden, 1976.)

H. Ornauer, H. Wiberg, A. Siciński, and J. Galtung (eds.), *Images of the World in the Year 2000.* (The Hague and Paris: Mouton, 1976.)

Erwin K. Scheuch, "Freiheit und Grenzen der Forschung." Mimeo, RIAS-Funkuniversität, (Berlin, May 1979.)

Spiros Simitis, "Datenschutz und Wissenschaftsfreiheit." In Max Kaase et al. (eds.), *Datenzugang und Datenschutz: Konsequenzen für die Forschung.* (Königstein/Ts., Athenäum, 1980.)

Chapter 16

Discussion of "Fear of Science Versus Trust in Science"

Summarized by Andrei S. Markovits and Karl W. Deutsch

Max Kaase began the discussion by adding to his paper some perspectives not immediately related to public opinion. The first aspect of the fear-versus-trust phenomenon he commented upon was "the question of attitudes." Looking back into "the brief history of public opinion research," he noted that it came as a tremendous shock very early in social science study of the subject that "mass publics were really quite ignorant." This, of course, had great consequences for the shaping of democratic theory, especially in the United States. A proposition that had recurred over the course of the conference, Kaase noted, was that public opinion was very volatile, and that, basically, the general public was uninformed and irresponsible. There was no question that the results of empirical social science research had shown that attitudes vary with regard to their crystallization and consistency. (A well-known article by Converse, he added, had coined the phrase "nonattitudes" in reference to an aspect of this phenomenon.) We know, Kaase said, that crystallization decreases as we move away from personal experience and from the immediate relevance of an issue to the individual.

One aspect of this subject, referred to earlier in the conference by Claus Offe, was the effect on the public of geometrically increasing information in our society, partially transmitted through the mass communication system. There is obviously no way, Kaase held, that people can cope with this increased knowledge as they once did; thus, he posed the question of what modern coping strategies have been adopted. One strategy is the ideologization of individual belief systems, in a sense resulting in the "clear reduction in richness of content, a streamlining of belief and knowledge." Research in the United States has pointed to an increase in the crystallization of attitudes along these lines, which may lead, among members of mass publics, to a higher ability to conceptualize. Another strategy suggested by Kaase was "to economize, to take in, selectively, only certain types of information." As a means of coping with increasing information, such a strategy leads to the shutting out of some areas of knowledge entirely, but Kaase would hypothesize that it may be the only means possible.

In the traditional dichotomy of mass and elite publics, Kaase said, it is difficult but possible to describe elites as high conceptualizers. But if one looks closer, one finds that elites are competent, in this context, only in their own specific area of expertise; in others, they may be as ignorant as the mass publics. Moreover, he said, we cannot speak of only one public, but must take into account a variety of publics, defined by various criteria that are broadly cross-cutting. In terms of information publics, with their diversity and variety, scientists may belong to only some, and in these they are not by necessity better informed or more rational than other members.

Kaase noted that, whenever we speak of attitudes toward science in the past, we must be clear in our reference to specific mass publics. In today's democratic societies, there is one constituting principle that defines the public very clearly and uncontroversially: one man, one vote. Traditional public participation, within these systems, is electoral participation, but there is vast evidence that mass publics, defined in a democratic sense, are superseding this narrow definition. They seek to broaden access to decisionmaking in the form of "direct action" politics, which can bear relevance to the institutional makeup of scientific decisionmaking. The impact of the environmental movement, for example, may supersede normal institutionalized processes. Kaase thus recommended that we differentiate between mass publics within the context of attitudes toward science in contemporary societies, making clear whether we are speaking of democratic publics in the sense of one man, one vote, or in some other sense. Finally, agreeing with an earlier contention by Joseph Weizenbaum that the participants should disclose their normative biases, Kaase said that he wrote his paper from a position holding to the desirability of free science: that is, science not dominated politically.

Jean Stoetzel was the first to comment on Kaase's presentation, adding some "explanations and information which should only complement this excellent presentation." The study of attitudes toward the year

2000, he said, was the first attempt by Eastern and Western European social scientists to cooperate in survey research, undertaken more as an experiment in East-West cooperation than as a substantive advance in the field. Its form thus outweighed its substantive findings. Stoetzel participated in this unprecedented cooperative survey project.

The first report of the study, he said, found that very few respondents in either the East or the West opted for the "Don't Know" choice on science- and technology-related matters. More surprising was the fact that there was a general consensus on these subjects between East and West, with only two countries at variance with the majority. (Italy showed far more optimism, and the Federal Republic of Germany showed more pessimism.) Many respondents said they would include psychology and psychiatry among recognized sciences. Other scientific areas did not lag far behind these two in general popularity.

Stoetzel also reported on the results of the second report of the study, giving details concerning public involvement in decisions about scientific policy. Forty-three percent of the respondents said that they were not in a position to discuss science competently, and 19 percent said that they had competence; nevertheless a general desire to be included somehow in decisionmaking concerning science was reflected in the survey. (In the portion that expressed this desire was a majority of the better-educated sample.) The young were the least able to perceive changes brought about by science since 1950, and the most skeptical of scientific advance were those who were either very old or who were educated only through the elementary school level. Stoetzel judged these results to be a very important contribution to better understanding of cross-cultural attitudes, on a mass level, toward science and technology.

Joseph Weizenbaum expressed surprise that Kaase would say that people's lives have become more predictable. Weizenbaum, in his interaction with young people, had reached an entirely different perception: he detected a growing malaise and a greater sense of uncertainty, conditions which would lead one to believe that—at least subjectively—people's lives have become less, rather than more, predictable.

Weizenbaum's next question was prefaced by his admission that, though it was widely accepted that such conferences as this one never taught anyone anything, he had learned from Karl Deutsch the importance of asking quantitative questions, even though it was not Weizenbaum's style to do so. Weizenbaum wanted to know from Kaase some of the particulars concerning the supposed increase in external controls over science. Was there really an increase? How could it be measured? Were the controls multiplying or declining? Weizenbaum also posed questions of a qualitative nature. Who controls science, what means are employed, and what are the purposes of control? Who gains and who loses? These bear at least equal, if not greater relevance to issues of control than do the questions of magnitude. Finally, Weizenbaum challenged Kaase's statement, made toward the end of his presentation, that scientists *had to* resist external pressure; why did he say *had to*? Was it axiomatic? Was it a law of nature?

Kaase began his replies by stating that life was more predictable to a greater number of people today than ever before. Because of the system of *"das Netz der sozialen Sicherheit"*—the network of social security and social services—people were better assured that, whatever the circumstances of, or changes in, the job market, there would always be something to help them cope; the system would not let them perish. This is the major contribution of the welfare state.

Kaase mentioned an article by Lenk and Ropohl that surveyed the longitudinal pattern of change in public attitudes toward nuclear power. The public began with optimism but has become increasingly skeptical during the last eight years. This, Kaase believed, reflected a very reasonable attitude on the part of the public as it reacted to new data and information. He added, in response to Weizenbaum's last question, that a free science, with fewer external controls and with greater responsiveness to internal institutions, was inherently more desirable. If one wished to label this axiomatic, one could.

Kaase conceded that his judgments of the increase in external control over science were not borne out by any quantifiable research that he had done. His opinion was based rather on his personal interpretation of singular evidence that he had witnessed and collected in an unsystematic fashion. One indicator was the amount of money channeled by the government into science. More important, as Harvey Brooks noted in his paper, was the very high distribution of scientists in government-funded and industrial research, a percentage which implies at least some control by these institutions. Kaase concluded that it would be fascinating, albeit difficult, to quantify this phenomenon and to attempt the necessary systematic approaches.

Klaus Traube then asked if the data bank concerning modern attitudes toward science was indeed as slim as Kaase had said it was, and whether questions on progress and its relationship to public opinion could not also shed light on the issue of science, a much less surveyed area. In this context, Traube recalled a survey conducted by *Stern*, the German magazine, which examined attitudes toward progress in 1972 and then again in 1978. During that time, Traube said, there was reflected a tremendous change in public attitudes toward progress. A United States survey similarly showed ambiguity concerning the notion of progress, an ambiguity which increased dramatically over roughly the same period surveyed by the *Stern* study. The United States study asked, among other things, whether or not progress had improved the quality of life. Though many of the respondents answered that life was better in the time of their ancestors, they nonetheless indicated that they prefer their own time.

Traube also questioned Kaase's assertions about the perceived predictability of life today. "If you say that life is more predictable," he posed, "what is, at the psychological level, this feeling of predictability?" Forty or fifty years ago, he said, Germany was mainly a rural country, with its people touched only peripherally by industry; people felt a continuity. They had few options complicating their lives. They had a lower life expectancy, but they knew what life would entail. On the other hand, industrial and industrializing nations all show a high rate of neurosis and suicide. Thus, Traube fully agreed with Weizenbaum that contemporary life has become less, rather than more, predictable—and not only for young people.

Elisabeth Helander, reacting to the point in Kaase's presentation that the public's comparatively low level of specific knowledge about scientific activities may be a determinant of public attitudes, raised the question of the links between public attitudes toward science and the structure of control of science. She believed that it is important to ascertain how public attitudes could lead to, on the one hand, changes in science's internal structures for decisionmaking, and, on the other, the nature of external control mechanisms and their relative social acceptability. External control was quickly growing; the pure alternative, however, would be science in a vacuum, with no societal connection, and with an accompanying danger that science would have no practical applicability. She proposed that research be conducted on the nature of new decisionmaking structures between scientists and the business interests that employ them.

Peter Weingart commented first on the nature of science coverage in the mass media. He noted one study that showed that science reporting in newspapers, in both the United States and in the Federal Republic, had risen from 2 percent before World War II to 8 percent in 1960. (No more recent data were available.) However impressive the increase, Weingart believed that the public deserved more and better reporting. He concurred with Kaase's earlier remarks about the nature of the term *the public* in a "one man, one vote" democracy, noting that the difficulty with polls is that though there is never one single public, the nature of our democracy forces politicians to pay attention to "the" public as if it *were* single and not diverse.

Moreover, polls have a further weakness. Worded in positive terms and vague generalities, they often elicit positive responses; as the terms get more specific, however, responses change. The word *science* in the public mind might be associated with medicine and not, for example, with the atomic bomb; in addition, the word is often inaccurately applied to technology, which is different from science. Furthermore, general or vague awareness of a scientific issue may foster certain attitudes, while recent knowledge of specific events may change these attitudes. For instance, one survey showed that awareness of nuclear power in the United States had increased from 77 percent in 1976 to 90 percent in 1979; after the Three Mile Island incident, the proportion in favor of nuclear power decreased by 14 percent, and the proportion against it increased by 23 percent.

Kaase answered Weingart that events are indeed important; he agreed that they influence science as well as secular trends in general. In response to Traube, who commented on the insufficiency of the data base concerning public attitudes toward science, Kaase noted that up to this point, the area of public attitudes toward science had not been sufficiently conceptualized and therefore not adequately researched. "There are certainly relationships between attitudes toward science and attitudes toward progress," he said, "but we do not know precisely the nature of this relationship." He agreed that the data base is indeed restricted, and noted that this conference had stimulated him to consider undertaking research along these lines.

On the issue of the predictability of life today as compared to life in previous eras, Kaase maintained that, in his paper, he used the term only in relation to the phenomenon he was discussing, namely, fear. The inclusion of the relationship of predictability to social circumstances, as had been done in the discussion, altered the nature of the question, turning it into an entirely different issue not immediately germane to his own paper.

In response to Helander's comment, Kaase reaffirmed his belief in the need to maintain science as a free institution; she had addressed the core of his reflections on the institutional makeup of science and the ramifications of external control. With increasing external control the pressure is greater now than ever before to implement internal controls, though this will require creative thought among scientists, since no precedents for such controls exist, at least within the social sciences in Germany, to Kaase's knowledge.

Kaase concluded by stating that he wanted to be counted among those who are rather optimistic about the ability of the mass public to react to the emergence of science's and technology's problems for society with maturity, insight, and dignity. On that note, the discussion ended.

Chapter 17

Fear and the Rhetoric of Confidence in Science

Harrison C. White

Confidence is . . . the opposite of what causes fear; it is, therefore, the expectation associated with a mental picture of the nearness of what keeps us safe and the absence or remoteness of what is terrible.

—Aristotle, *Rhetoric* (1383 a 15)

Forces exposing science to fear and distrust are not as remote from science itself as it would be convenient to believe. Within a science, the same achievements that lead to more advanced baselines appear as temptations to turn from further advance in order to spin elaborate webs of restatement, taxonomy, and the persuasiveness of the new truths—in short, to embrace a rhetoric of confidence. Rhetoric is without doubt as old as science, and continues as its chief competitor. Each science requires a self-conscious discourse in which persuasion about definite features of the natural world is a distinct goal. Rhetoric, in contrast, "is not concerned with any special or definite class of subjects" (Aristotle, 1355 b 35).

Science (the sciences as a whole) also is tempted toward a rhetoric of confidence in its efforts to connect with society, especially through

Note: Support under NSF Grants SOC76-24512, SOC76-24394, and SER76-17502 is gratefully acknowledged. I am indebted both to Robert K. Merton and to Ronald L. Breiger for correcting some errors and for bringing out latent ideas in earlier drafts.

applications provided in return for financial support and social recognition. Scientists would be naive to think that they controlled interaction across these boundaries, which are patrolled by intermediate layers of elites, officials, and public interpreters, often with causes of their own to clothe in mantles from science and the sciences. The structure of this situation is dangerous to science and is fraught with feedbacks and possibilities for manipulation. Not least perilous is the likelihood that a rhetoric of confidence espoused by scientists and their spokesmen will lead to fear and to the erosion of trust from society at large.

Forty years ago, the American sociologist Robert Merton probed two orders of reasons for the revolts he saw then against science (reprinted 1957, p. 538). One was the conflict of results or methods of science with important values; the other was a sheer emotional incompatability in ethos. One reading of his text suggests that important revolt was confined then to some nasty societies, totalitarian ones; our own liberal societies bore no such incubus. And yet, consider the extraordinary words of Robert E. Kohler, a historian of science, in a recent letter to *Science* (29 February 1980, p. 935): "Historians and sociologists of science must contribute to an honest and realistic picture of the scientific enterprise as a social institution, not different in any fundamental way from other economic, cultural, or political institutions." Surely this is a rhetorical goal, not a scientific one. There may be an irony in Merton's having been a founder of the sociology of science, if I am correct in seeing him and his leading students as sturdy admirers of sciences as they actually are, as well as of Science.

This line of thought is too dramatic, too extreme, just what one must fear when asking about Science with a capital S. Let us return to my sober view that fear/distrust, and thus the effectiveness of sciences in our society, depends a great deal on changes in the sciences themselves.

Diverse variations come to mind on my basic theme, which is *maturing sciences adopting rhetorics of confidence:*

1. the social sciences poised between reflexivity and natural sciences;
2. the Latinization of mathematics;
3. Aristotelian sciences reemerging as modern statistical "science";
4. the humanities searching for a new lease on life by adhesion to science, or rather by capturing some commanding heights of science;
5. a joint struggle of arts (unsuccessful so far) and sciences (successful up to now) against the horde of commentators, the Academy crowd;
6. science moving from discovery (miracle) to cosmetics of presentation;
7. increasing gaps between sciences and technologies.

Common to these variations is my emphasis upon distinctions among particular sciences as well as upon their differing interfaces with other forms of knowledge.

A curious change in past decades is the lack of controversy over incorporating certain new fields as sciences. There is even a new area for Nobel prizes in science, and some of these awards have been given to a new breed—persons whose contributions could not be overturned by any conceivable data. Is there perhaps, after all, some strong pressure toward change in science coming from our society quite independent of internal tendencies in science? It is plausible. As other forms of legitimation lose even their twilight afterglow, Science becomes, if only by default, much more important for general ideology, while the sciences become targets for blatant manipulation by particular institutions and elites. If so, one should expect any general profile of fear/distrust in society to carry over toward science, as people begin to sense this ideological cooptation, this subtle reliance on science as the last confident rhetoric of confidence.

According to Merton (1957, Ch. XVIII), science in England was especially closely tied to Puritanism in the seventeenth century. Science was justified by other values and views, which were carried on a vigorous new wave in religion associated with rising social strata, if not yet a new class. (See also his preface to the 1970 reissue, available also in Storer, 1973.) If I can perceive a vigorous new theology associated with possibly new social formations, I might use the analogy as a clue to predicting *improving* trust today for sciences on the part of some publics, a case of science sheltering behind *other* institutions. Possibly I could argue that on the other side of the Berlin Wall there is a glorification of science, rather consciously promoted in the name of a rising class. But I would be chagrined to note, just there, hostilities to emerging new sciences, which are to me the healthy growing edge of science as an institution.

I turned back to reread Perry Miller (1967) on Puritanism in the provincial stronghold where I work. Miller's volumes are entitled *The New England Mind* justly; his preoccupation with doctrine is total, and so, according to him, is the reach of Puritan doctrine across the world of ideas. He shows very clearly (Ch. VIII) that the complaisance of Puritan theology to the new science extended equally to Peripatetic physics (Miller, pp. 216, 224, 235). The opportunistic flexibility of Harvard's views of science under Morton, the vice-president brought over from England (Miller, p. 221), suggests little reliable support for science from the more worldly part of the Establishment, Puritan or no. Merton perhaps slights doctrine and New England, and Miller certainly slights class, organization context, and England; but between them the coverage of Puritanism and Science is superb. From Merton as from Miller it is hard to argue any powerful motivation from Puritan religion to carry forward what we now know as Science; it is easy to argue with them that Science then as now was regarded by others as a convenient rhetoric of confidence. What support there was then for science in its own right had very much the aura of our own day's Unity of Science movement, which in my opinion so often is a cover for hostility to actual sciences at their tangible, erratic growing edges.

If we resume examining the list of seven variations on my theme, it becomes clear there is no such thing as a pure internal change within the sciences, one not "contaminated" by the increasing pressures from the society. Consider the second variation, the changing place of mathematics, which crosscuts several of the others.

A case can be made for the Latinization of mathematics. Mathematics has been an enormously productive and powerful servant of science; it has helped discover and master new, concrete phenomena. Does one now see more and more a different use of mathematics, its use as mystification, as a mere translation into a high language, the new language for a rhetoric of confidence? (cf. Weyl, 1949, p. 235.) A possible signal of this is the increasing tendency to view mathematics as self-contained or even as above the sciences, their very queen.

A sign of growing isolation of mathematics from sciences is the growing preoccupation of mathematicians with the "foundations" of mathematics as such (Snapper, 1979). In the nineteenth century there finally expired the last vestiges of the medieval Trivium, the leavings from Aristotle's bold vision of rhetoric (Miller, 1967; Ricoeur, 1975). Could it be that mathematics, which was in the competing Quadrivium, now in this century is being groomed to take rhetoric's place? To me, two of the three foundations proposed for mathematics (Snapper, 1979), Logicism and Formalism, carry the aura of rhetoric, and it is the third, Intuitionism, which is most widely rejected by mathematicians.

At one level rhetoric adds force to deep arguments (Kurosh, 1965; MacLane, 1972). At another level rhetoric is the fashionable way of selling one's nostrums. Is it really a deep love of science together with its commitment to truth, that has been leading every school of business and every school of policy rhetoric to mathematics? Can the genuine scientific use of mathematics be unaffected? Can attitudes toward mathematical science on the part of our fellow citizens be unaffected? (Talking about rhetoric leads me into rhetorical questions!)

I am much exercised by such questions, since my trade is mathematical models in sociology. I find the issues slippery and two-faced. Within mathematics itself the long-term trend toward elegant structuralism (MacLane, 1972) abets use of mathematics as a rhetoric of confidence. But the noted mathematician, Hermann Weyl, who, as I mentioned earlier, warned of this dangerous slipping away of mathematics from science, in the same work emphasizes the scientific value of the new structural types of mathematics.

Combinatorics is an essential mathematical base for social and for biological theory. As Weyl makes vivid, classical analysis, on the other hand, is profoundly intertwined with the specific physical properties of space. Social and perhaps biological processes may be little affected by physical space and much affected by other sorts of combinatorial constraints in the environment. Substantive scientific uses of combinatorial and algebraic mathematics are hard to disentangle from rhetorical use, just as it is in fact difficult to sharply distinguish "combinatorial" or "structural" mathematics from "classical analysis."

Let me outline a clear illustration of all these confoundings at once. René Thom is the preeminent example, in every sense, of the mathematician bringing a structural rhetoric to the world of affairs. His catastrophe theory surely is well known, even to those who have not followed Christopher Zeeman's energetic attempts to turn it into bread-and-butter applications (Poston and Stewart, 1978). Thom himself intends an intellectual breakthrough for biological and perhaps social sciences, a structuralist breakthrough. To date the overwhelming verdict from social and biological scientists, as far as I can see, is disdain (e.g., Berlinski, 1978, but see Fararo, 1978). Yet, in complete independence—indeed I should say innocence—and over just the same period, the early 1970s, Kenneth Wilson (1979) developed a major new approach to the physics of matter which to me shows striking similarities to Thom's line of work. This and other similarities to results of *physical* science are made the central validations of catastrophe theory by the young Englishmen Poston and Stewart: from a breakthrough in biology to a pedagogical aid in physics!

Wilson's "renormalization group" is an adaptation of quantum electrodynamic technique to phase transitions in matter (Nelson, 1977). Many physicists were and are a bit nonplused by this extraordinary crossbreed. They are, if you like, shaken up by exactly the Thom-ist flavor. They are shaken up by the mix between tangible structure, real space, and abstract "structuralism." It is difficult for physicists to deny that this work is a major contribution to the theory of cooperative phenomena. Yet there are two rubs. One is that the Thom and Wilson sort of theories are a gift from heaven to rhetoricians of almost any socio-economic-political persuasion. The other is that Wilson is able to explain so much of his chosen phenomena (of turbulent flows and critical points in phase transitions) exactly because he is in a way proving how little regularity there is, how little lawfulness exists in these phenomena according to his theory.

A remark by Paul Ricoeur in his brilliant work on metaphor (1975,

p. 12) suggests to me an alternative successor to rhetoric from the sciences.

> The great merit of Aristotle was in developing this link between the rhetorical concept of persuasion and the logical concept of the probable, and in constructing the whole edifice of a philosophy of rhetoric on this relationship.

Aha! Maybe it is statistics, whose practitioners increasingly push it forward as a science, which is best suited for taking over the place of rhetoric. To use statistics has always had an essentially Aristotelian flavor, which is to say an aura of common sense, while mathematics, like science, has in everyone's eyes a more mysterious quality. In a recent critical examination of statistical inference as actually carried out, the author says (Leamer, 1978, p. 285): "In fact . . . a strong argument can be made that statistical inference, not Sherlock Holmes inference, is unscientific." And earlier he says (p. 13), "There is a growing cynicism among economists toward empirical work. Regression equations are regarded by many as merely *stylistic devices*" [italics added]. These signs do point toward statistics as the new rhetoric of confidence.

Statistics relates much more naturally than structuralist mathematics to the mathematization of "policy sciences" on which I remarked earlier. And often the policy use of statistics per se is intertwined with the policy use of mathematics. Take just one recent example, from operations research on the optimal distribution of fire companies in New York City (Chaiken and Larson, 1972; Kolesar and Walker, 1972). This work elicited a vitriolic denunciation from a science-for-the-public grouplet (Wallace and Wallace, 1977), which argued that the analysis relied on naive mathematical formulation and poor statistics, thereby functioning as a screen for budget slashes and abandonment of huge stretches of the city to firestorms. If true, this is a stunning indictment of statistical science as a rhetoric of confidence. The charge of naiveté is hard to maintain in face of later explications (Larson, 1979), and strong rebuttals of these particular charges are to appear.

These examinations of the second and third variations of my theme yield ambivalent results. They prove representative of my final conclusion: The signs and avenues of growth and change in the spectrum of sciences are simultaneously signs and avenues of manipulation, and of general exploitation for a rhetoric of confidence.

I myself wish for a move toward less fear and more trust of science only on two conditions. One is that the sciences remain a relatively free and independent institution. (The problem here, to me, is not that basic science will be suppressed but rather that it will be drowned in a flood of manipulative rhetoricians.) The other condition is that science continues to produce miracles.

Make no mistake. Miracles were required of religion for it to become established. Science was accepted with fewer missionaries (and less conversion of royalty), because people could see its miracles more frequently. Will—indeed, should—people continue to be as impressed?

On the one hand, I argue, there is an increasing humanitization of the sciences. They come to resemble humanities more in a number of aspects: The creators of the field are required to be dead so as not to interfere with the labors of their inheritors, the humanistic scientist-scholars, in their task of polishing a rhetoric. The unpleasant ebullience of genuine discovery is replaced by the charm of scholarly ping-pong adapted from traditional Chinese literati. Surely one reason for the enormous resonance of Thomas Kuhn's book, *The Structure of Scientific Revolutions,* is his having reported, and thereby no doubt also having encouraged, the possibilities for turning any science into a humanity (cf. Blaug, 1975, pp. 403–405).

Ricoeur (1975) has recently elaborated Max Black's case for this resemblance of science to a humanity. Any scientist will quail at the embrace when phrased as James Ackerman does (in Holton, 1957, pp. 18–19): "We shall never come together in the laboratory or among the philological footnotes; only in the atmosphere of *scientia* shall we find the exceptional fraternity of creative processes that constitutes a cultural communication center, ... The significant distinction in our culture, then, is not between the arts and the sciences, but between the limited technicians and the creators of images." Harry Levin (in Holton, pp. 3, 9) is harsher with scientists:

> Beneath its [C. P. Snow's Rede lecture] well-meaning truisms there lurks one striking novelty, which has scarcely been tested by all the discussion: the implication that science can stand by itself as a culture. This assumes not only a total separation from the humanities but an internal unity among the scientists . . . . It seems to be an intrinsic feature of the situation that most of the dialogue between the sciences and the humanities must take place within the latter's domain.

This first trend portends hostility toward the genuine changes I mean when I say miracles. And in older sciences the genuine changes become harder and harder for outsiders to perceive. Glittering changes, from the reshuffles at which humanities excel, become easier. Ackerman is closer to the truth than Levin: science now can stand free as a rhetoric.

On the other hand, sciences, even centuries after their coming of age, are endlessly rejuvenated by whole new fields. Astrophysics is one such, by my reckoning. One by one, the once ridiculous *dei ex machinae* of E. E. "Doc" Smith's Galactic Lensman series are made plausible by new concepts brought into the science literature by John Archibald Wheeler (in Seeger and Cohen, eds., 1974, pp. 257 ff.), among others. On the technological side, Wheeler's Princeton physics colleague O'Neill (1977) urges immediate launching of facilities to mine the moon and thence to create orbital "factories" to build the huge fleet of solar power stations we need.

Here we come to a paradox I see in the application of science. The process of maturing, even without any imposed tendency to rhetoric for its own sake, tends to complicate the lines from science to technological imagination and its realization. But in practice a more important blockage is a spinoff of science itself. This spinoff was described earlier: Mathematized rhetorical gloss for decisions—adopted to garner legitimacy from science—is, I think, likely to block any genuinely bold technological change.

There is, I think, an analogy here to the effects of the change from corrupt to "impersonal" civil service in the United Kingdom (cf. Kelsall, 1955). An erratic, personalized favoritism, which could admit major change and recognize completely fresh talent, was exchanged by the famous North "reform" of 1870 for a system limited by an explicit rationale. The class bias in its concrete operation was matched only by its closure to change: a classic case of the rhetoric of confidence. I could introduce formal arguments, but to go far in that direction is to accept the rhetoric of policy science models themselves. Surely the sheer silliness of judging O'Neill's (1977) solar power factories using current neoclassical economic arguments and measures is evident.

It is a bit sobering as I look back over these notes and crochets to recognize how many of them can be subsumed under headings in Merton's 1937 essay. And there is one more parallel, to his "Public Hostility Toward Organized Skepticism." You will not be surprised to hear that I, a sociologist, think social sciences may score the most miracles in the next decades. I think, however, these can come about only when our sciences have come to terms with reflexivity much more fully than we have managed so far. It is hard to imagine this development not producing the sharpest objections from our society.

You see that I have strayed from my initial charge for this conference. I have asked not so much how common people's attitudes toward science will be changing, but what new pressures and invasions of science by other institutions and elites can be anticipated. But then Karl Deutsch and Shepard Stone as conference organizers would, I am sure, agree that a distinctive hallmark of science is the freedom, indeed the impelling urge, to seek to uncover the question most in need of asking.

They still can rightfully ask: Overall, what is the drift on fear/trust, and what is there to be done? My main guesses, after surveying seven variations on my theme, are:

1. science has become the only ideology with legitimacy across many entire societies;
2. this legitimacy is thin and weak with most persons, in spite of its universality;
3. tides in the fear/trust spectrum will depend—aside from specific technical areas such as nuclear power and biotechnology—on general swings of mood about one's society, modulated by the prominence its elites have been awarding to science;
4. material support from elites for science will continue largely unaffected by the tides in public mood, partly because the cost is so low (notably in the sharing of power, where few of science's leaders have shown the hubris traditional for lords spiritual);
5. the sciences will have a difficult time policing and protecting themselves from enhanced exploitation: not just from specific manipulation but from abuse as a general rhetoric of confidence.

I should like to end with a particular concern: How are views of science evolving on the part of that very large body of persons known as women? Banality is the best I can offer on a general level. As persons who are women become known as persons of other sorts, the question will/would evaporate. According to *Fair Science,* a broad-gauged survey just being published, differences already have nearly evaporated between men and women in how they are judged while functioning within science—or, to be precise, Johnathan Cole adopts that as his null hypothesis and is not shaken from it by the data he searches out or by his fresh surveys.

I do wish to mention a specific sign of the changing position of women, within sciences as a culture. First there was Ursula le Guin, at last a woman doing science fiction openly, without a nom de plume. Karl Deutsch would, I am sure, agree that science fiction is one of the best possible, because the least self-conscious, indicators of mood toward science, by reader and by writer. Yet le Guin is unique, too advanced to prove the point. Now, finally, there are women writers who openly do bread-and-butter "sci fi." I particularly wanted to cite Nebula Award Winner Katherine MacLean, because I was so delighted to find in her work *Missing Man* (p. 24) resonances from my third variation on the theme of rhetorics of confidence in science.

... Ahmed stood up and shouted. "You're talking like a caveman ... I don't care what your reasons are; nobody cares anymore what the reasons are. We only care about results, understand? We don't know why things happen, but if everyone makes out good reports about them, with clear statistics, we can run the reports into the machines, and the machines will tell us exactly what is happening, and we can work with that, because they're facts, and it's the real world . . . . Scientific theories about the causes don't matter."

He was red in the face and shouting, as if I'd said something against his religion or something. "I wish we could get theories for some of it. But if the statistics say that something funny happens here and something else funny always happens over there next, we don't have to know how the two connect . . . ."

REFERENCES

Berlinski, David. "Review of Poston and Stewart," *Behavior Science,* 23, 1978, 402–416.

Blaug, Mark. "Kuhn versus Lakatos, or Paradigms versus Research Programmes in the History of Economics," *History of Political Economy,* 7 (1975), 399–433.

Chaiken, J. M., and Richard C. Larson. "Methods for Allocating Urban Emerging Units: A Survey," *Management Science,* 19, 110–130, 1972.

Cole, Johnathan. *Fair Science.* New York: Free Press, 1980.

Fararo, Thomas J. "An Introduction to Catastrophes," *Behavioral Science,* 23: 291–317, 1978.

Holton, Gerald (ed.). *Science and Culture.* Boston: Houghton Mifflin, 1967.

Kelsall, K. K. *Higher Civil Servants in Britain.* London: Routledge and Kegan Paul, 1955.

Kolesar, P., and W. E. Walker. "An Algorithm for the Dynamic Relocation of Fire Companies." The New York City Rand Institute, Research Report R-1023 (published subsequently in *Operations Research),* 1972.

Kurosh, A. G. *Lectures in General Algebra.* Oxford: Pergamon Press, 1965.

Lane, Michael, ed. *Structuralism: A Reader.* London: Johnathan Cape, 1970.

Larson, Richard C. "Structural System Models for Locational Decisions: An Example Using the Hypercube Queueing Model," in K. B. Haley (ed.), *OR'78.* Amsterdam: North Holland, 1054–1091, 1979.

Leamer, Edward E. *Specification Searches: Ad Hoc Inference with Nonexperimental Data.* New York: John Wiley, 1978.

MacLean, Katherine. *Missing Man.* Berkeley: Medallion, 1976.

MacLane, Saunders. *Categories for the Working Mathematician.* New York: Springer-Verlag, 1972.

Merton, Robert K. *Social Theory and Social Structure,* rev. ed. New York: Free Press, 1957.

Miller, Perry. *The New England Mind: The 17th Century.* Cambridge, Mass.: Harvard University Press, 1967.

Nelson, David R. "Recent Developments in Phase Transitions and Critical Phenomena," *Nature,* 269, 379–83, Sept. 29, 1977.

O'Neill, Gerard K. *The High Frontier.* New York: W. Morrow, 1977.

Poston, T., and I. Stewart, *Catastrophe Theory and Its Applications.* London: Pitman Publishing Ltd., 1978.

Ricoeur, Paul. *The Rule of Metaphor,* trans. Robert Czerny. Toronto: University of Toronto Press, 1975.

Seeger, R. J., and Robert S. Cohen, *Philosophical Foundations of Science.* Hingham, Mass.: D. Reidel, 1974.

Snapper, Ernst. "The Three Crises in Mathematics: Logicism, Intuitionism and Formalism," *Mathematics Magazine,* 52, 207–216, 1979.

Storer, Norman (ed.). *Robert K. Merton: The Sociology of Science.* Chicago: University of Chicago Press, 1973.

Wallace, Rodrick, and Deborah Wallace. *Studies on the Collapse of Fire Service in New York City 1972–1976: The Impact of Pseudoscience in Public Policy.* Washington, D.C.: University Press of America–R. F. Publishing Inc., 1977.

Weyl, Hermann. *Philosophy of Mathematics and Natural Science.* Princeton, N.J.: Princeton University Press, 1949.

Wilson, Kenneth G. "Problems in Physics with Many Scales of Lengths," *Scientific American,* 158–179, Aug. 1979.

Chapter 18

Discussion of "Fear and the Rhetoric of Confidence in Science"

Summarized by Andrei S. Markovits and Karl W. Deutsch

Harrison White began the discussion by adding to his paper some concrete examples of trends that had already been broadly outlined over the course of the conference. Without denying the importance of the ideological issues of control or the nature of the technological spinoff aspects, he wanted to present a somewhat different point of view. In effect, he said, he wanted to take Claus Offe's statement concerning the deficit of control complexity as against design complexity and give it an ironic twist. He chose to focus on a central feature of science: its intrinsic intellectual content, and issues within the development of science itself that contribute to "this worrisome loss of control."

His concern, he said, could be portrayed in a phrase: "the increasing movement of science into rhetoric": not rhetoric in the pejorative sense in which the word is used today, but in the classic Aristotelian sense. This kind of rhetoric, he stressed, can be a beautiful thing, but it is not science. To make his point, he pictured "a three-layer cake: there is science, there is society, and I am claiming, there is a layer of icing, a very delicious but a very two-faced layer, which is rhetoric." This has been developed by science itself in many ways, and has become a subtle vehicle for scientists to distort their work before the public, and certainly for the more sophisticated part of society to control science.

Before developing this idea, White pointed to the first pitfall, which was that the word *science* itself was taken for granted, because the conference participants were so involved in it. His distinction was that the word *science* was most often used, and not *sciences,* implying that the participants had forgotton the differences between, for example, social science, biological science, and natural science, along with the implications of such divisions. White "warned" the conference members to remember the difference, and "remind[ed them] that the world knows the difference." He pointed to the fact that, during consideration by the United States House of Representatives of the budget appropriation for the National Science Foundation, a complete excision of funds for social science research had been approved. News of this never reached the newspapers because the Senate restored the funding and "it all came out right in the end"; nonetheless, the measure had been debated and approved by the House. In the political realm, there was a crucial awareness of the divisions within science, and of their respective importance to the priorities of society.

White's main point, however, was the shift of science to rhetoric. His theme was "The Latinization of Mathematics?" He wondered whether mathematics had taken on the nature of *Hochdeutsch,* if it was becoming a fancy language, developed without the intention but with the effect of hiding unconvincing, unscientific, or shallow analysis behind a literary layer. He conceded that, since the use of mathematics in social science was his life's work, he obviously did not believe it to be an unmitigated evil, but that, in order to remain consistent with his present rhetorical device, he would say nothing more in its favor.

White asked, "What about mathematics itself?" He noted that one can argue that there has been an extraordinary development in the "librarianship" quality of mathematics, that it had become imaginative and inventive, but that, with the emphasis on indirect proofs, on elegance, on classification, mathematics was taking on all the signs of a high and artificial language. As an example he cited René Thom and his catastrophe theory, a scientist bringing rhetoric into the public sphere, "quite self-consciously, as he will tell you himself." This mathematical theory was greeted with disdain in biological and social science journals, White noted, and thus "it would seem that I have an easy victory, that, yes, there is a Latinization of mathematics." But he also noted that a new book by two young English physicists brought out many of the same aspects of catastrophe theory, "this apparently rhetorical device," within the field of phase transition in physics. An ironic twist ensues, he said, because this development in physics was a negative one; the power of the new theory was that it reached the point very close to the limits of chaos, yet one could get very powerful mathematical results of great generality.

The real content of the formal results was that we cannot know very much. He mentioned this example, he said, in order to be fair on the Latinization issue, suggesting that, although Latinization may have worrisome implications for social control, it may have genuine connections to the nature of the progress of certain fields of science.

White presented statistical theory as an alternate candidate to replace mathematics as the prime embodiment of Aristotelian rhetoric. One interesting aspect was the passivity inherent to statistical theory, since the world was always in the protected position of being the null hypothesis. The world as the null hypothesis, White said, smoothes out statistical theory, adding to its rhetorical attractiveness.

He introduced a final point by referring to the book, *Studies on the Collapse of Fire Service in New York City 1972–1976,* subtitled *The Impact of Pseudoscience in Public Policy.* It was an example, he said, of the linkages between mathematics and the "so-called" policy sciences; its thesis rested on a critique of a Rand Institute operations-analysis study on firefighting in New York City. The purpose of the book was for "outlaw priests" like its authors to defrock and outwit the bishops of the scientific establishment. It is precisely for this reason that the authors of the volume and others like them have had to make do with shoestring budgets while established scientists remain very comfortable.

The reason, White said, that the whole complex was so hard to evaluate—partly tying back to catastrophe theory—was that the fundamental point of the authors' study argued that fires in New York City should be treated as an epidemic situation, handled with the kinds of models appropriate to stochastic epidemics. Rand, on the other hand, saw the problem to be one of simple operations research. The literature discussed complicated problems and technical issues without a common matrix: operations researchers had no idea what stochastic epidemics were, and most of the biologically trained stochastic epidemic researchers were not amenable to the framework of operations research.

White closed with a thought on the nature of the fear/trust problem. Since he himself believed the public to be, on the whole and in fundamental matters, realistic and in possession of good judgment, he was inclined to look critically at science to see if maybe the public had sensed something that science had not. If fear escalates, he proposed, maybe science should devote itself to a measure of introspection, in order to perhaps discover the causes. It was not the public that worried White, but science itself.

Hans Wolfgang Levi expressed the view that, while most of those present believed that public distrust was the prevailing attitude toward science and that control of science is a way of helping to overcome this distrust, he did not think that greater public control was the answer.

Through the mass media, the public already interferes with science, and the effects of such interference could be disastrous. Free exchange of views among scientists could be hampered by public attention; already there was a trend toward hiding controversial results to avoid public uproar.

Furthermore, he asked, how does one control science? Scientific research can only be controlled by money, and even this control applies only to expensive research. Moreover, whose standards would be used in future control? Who will decide which research can be permitted and which cannot? Any kind of external control over science, he said, was intrinsically shortsighted.

Gerhard Leminsky addressed the subject of modes of public participation in science by noting that, while Karl Deutsch had said that the participants were seeking a common orientation toward the problems of science, such a universal approach could not be achieved by taking into account only the viewpoints of those present at the conference. A common orientation, he said, presupposed a common landscape, and the participants were from only one landscape, one background: by and large, they came from reknowned and established research institutions and thus reflected the concerns of "big science": atomic energy and genetic engineering, for example. The landscape of the large majority of working people, however, had been woefully neglected; working people have a different point of view than scientists, and indeed their whole *Problemstellung* would differ from the framework presupposed at the conference.

The concerns of working people revolved around such issues as the quality of work life, training and education, and the creation of a humane environment. These concerns had been treated only superficially at the conference. Perspectives, he reminded, determine the definition of problems. Most participants had spoken of human adaptation to technologies such as computers, as if humans were the slaves of technology. In industrial settings, adaptation to assembly lines, machines, and rationalized work organization had been a fact of life for decades. As an example of how greatly perspectives of a problem could differ, he noted that if we were to view a forty-five-year-old politician, manager, or scientist, we would call him a bright young man; on the other hand, a forty-five-year-old assembly line worker would be regarded as old and approaching the end of his usefulness. The conference had conducted discussions almost exclusively in terms of big problems and big perspectives, because the participants were members of the "big science" establishment; to consider all aspects of the daily application of technology, however, other perspectives must be included. If a worker has a good job, he is likely to look upon scientific developments with optimism. If he has a bad job, he is likely to be skeptical. Any consideration of social optimism or fear must be made within the context of people's daily lives.

If science is indeed an amplifier and not a modifier of society, as Deutsch proposed, then the structures of society and science, and the mechanisms between the two, should be discussed. Underlying the problems of control were problems of structure, and underlying problems of structure were differing conceptions of interests. Leminsky believed that the nature of control had not changed very much, but that the structural components of that control had. Social scientists seemed neither to understand this nor to care. In his own daily work in the unions, Leminsky had seen social scientists visit plants looking for specific problems along the lines of ergonomics or work organization; they failed to view the totality of worker concerns. This, in effect, meant that their research could not therefore contribute to any meaningful change, negating the entire purpose of their inquiry in the first place.

At one large factory, Leminsky said, the company had presented to the workers a new outline of work organization, designed to improve communication links, access to machines, and other points of efficiency. The workers rejected it, to the frustration of the engineers who designed the plan. The engineers, in their negligence of human concerns, had excluded human considerations, and had instead planned everything around the machines. This case paralleled to a large degree the earlier argument by Weizenbaum about the dangers and pretensions of "scientific imperialism."

"What can we do?" Leminsky asked. Scientists were thinking in theories, while workers were looking for answers to their daily problems. Sociologists study only segments of reality, which sometimes contradict one another, while workers were concerned with the totality of social problems. The solution lay with two approaches: problem-orientation and participation. Scientists needed to turn away from their models, their pure research, and their scientific journals, and focus instead on the solution of the problems of daily life. To shed their objectification, scientists needed to turn to the people for guidance, for feedback. People must participate in the solution of their problems, not just serve as objects or factors in research aimed at solutions. "If you are looking for the social implications of science," Leminsky said, "the scientist is the expert in methods, and the average person is the expert in the problems." If attitudes affect institutional control of science, and one wants to seek positive attitudes toward science, then the structure of science must be changed, for it is structure that determines attitudes.

Horst Ohnsorge referred back to a previous discussion in which it had been maintained that it was the use of science, and not science itself, that was dangerous. If this is the case, he said, then it is not the scientists who should be controlled, but those who make the decisions as to how to use science and scientists. Principally, this means politicians and industrialists. Scientists, if they suspect that their science can be put to dangerous uses, should convey this knowledge to those who decide on the social applications of science. In this way, scientists could exercise the power of their knowledge to ensure that their science will not be used in ways harmful to society.

Bruno Fritsch asked White if it was not one of his [White's] points that endogenous processes of science and scientific endeavors were of such a character that they surpassed by far any political or engineering control demands. White answered that this was roughly accurate, preferring to put it another way. For various reasons, he said, scientists had built a layer of rhetoric between science and the leaders of society, and that this layer had made it possible for outsiders to reach within the inner workings of science and manipulate it. But there was a danger equal to that of manipulation of the uses of science by "politicians sitting in another building somewhere," and that was the potential for manipulation by scientists themselves. The layer of rhetoric prevented anyone unfamiliar with the inner workings from ever knowing if science had been manipulated for one reason or another.

Klaus Traube expressed approval that Leminsky had raised the normative questions of just whom science should serve and what effect it has on the worker at the shop floor. The free communication between elites is not the main concern of the vast majority of people, Traube said, but what is more important to them is the flow of information between the public and the science that serves them. There is no valid purpose to scientific information that cannot be shared by those on whom it has its greatest effect. In this context, Traube would not mind if the public refused to cooperate with surveys unless it knew who paid for them and what their ultimate use would be. He said that it would do no damage to expose science to the public, because the supreme value should not be the progress of science itself, but the progress of the service that science can do for society. It is a normative question of public self-determination, he concluded.

In this connection, Karl Deutsch raised the point of ethics in medicine. Should doctors lie to their patients or tell them the truth? If a doctor told his patient the truth, Deutsch speculated, his patient might refuse an otherwise beneficial operation. Another possibility is that, if there is indeed such a thing as pyschosomatic disease, patients might be scared into the very disease of which the doctor was trying to cure them. One solution is Plato's noble lie—in this case, medical falsehood—but, in general, Deutsch suspected that the more physicians do this, the more they transform themselves into veterinarians, whose patients cannot talk back and have no share in the decision process. Deutsch felt closer to people who suggest that only in exceptional or marginal cases should doctors not tell the truth to their patients.

Deutsch then spoke directly to Leminsky's points, saying that there are two kinds of expertise: one was how to make shoes, the other was determining where the shoes pinch. If one does not ask people where the shoes pinch, then one will make miserable shoes no matter how expert a shoemaker one is. "Wise scientists who think that they will lie to the stupid public for its own benefit deprive themselves of the return flow of information, without which they cannot function, even with the most sophisticated techniques of survey research." If there is doubt about the utility of lying, Deutsch said, then do not lie.

Joseph Weizenbaum noted in this context that Americans had fought a revolution on the same principle: taxation without representation. In this case, the tax is not money, but, metaphorically, the intrusions of science upon the lives of people. Deutsch added to this point that, in terms of the "tax" or effect on people's lives, there are two ways of making a car—analogously, science—go faster: either by pressing the accelerator or by not stepping on the brake.

White interjected with an expression of concern, because many of the preceding remarks suggested to him that attacks on science were misplaced. Science, he said, was one of the few magnificent endeavors that societies have created, and he would hate to see it become "a whipping boy" for all the tensions and dissatisfactions that are grounded in some very fundamental aspects of the world and all social systems. His heart went out to Leminsky—whom he saw as a modern-day Jürgen Habermas in the flesh—but White did not think science was the one to blame for the problems Leminsky denoted. White suggested that these concerns might be brought to the attention of the German manufacturers' association, but pleaded that, whatever the case, they not be blamed on science, "this marvelous but feeble institution." His defense of science in this regard, he said, was the obverse of his previously stated concern that scientists might manipulate science.

Shepard Stone directed a question, concerning popular reactions to science, to Traube. How would Traube view the Scopes "Monkey Trial" in Tennessee of the 1920s, in which a schoolteacher was brought to trial for violating a state law against teaching evolution in the public schools? Here it was clear that the people of one part of the country at one point in time did not want something taught that the scientists believed to be essential to modern secular education.

Traube answered that what he had advocated was a dialectic process, a dialogue between scientists and the public. By this dialogue, he conceded, some truly beneficial effects of science might be slowed. He asked, however, if it was necessary that the beneficial effects come as soon as possible, under any circumstances, or if it was more important that there be a connection between what goes on in science and the understanding of it by the public. He believed that the latter, with its emphasis on participation, was of greater social value in the long run.

Leminsky concurred with Traube, adding that such a statement did not imply that scientific issues should be put before the public for a vote one way or another; it was more complex than that. The point is that science must accept people, with their needs and problems, not just view them as instruments. Dialogue was the essential imperative. Stone countered that, in the case of the Tennessee trial, people did not want a dialogue; they wanted to silence the teaching of a scientific concept. Weizenbaum, in turn, differed with Stone, injecting that the very act of a trial was an embodiment of the dialectic process of which Traube had spoken.

Deutsch reminded the participants of what Levi had said earlier, that (to generalize a principle of political science) if a weak group is oppressed, its prime weapon of self-defense is deception and conceal-ment. For instance, at one time, it was believed that diseases could only be cured by introducing good substances into the body, and that therefore the administering of an otherwise poisonous substance as a cure was in itself an evil thing to do. But many medical and pharmacological developments could not have occurred without the latter method; thus, at the time, it was carried out in secret. Similarly, anatomy was practiced secretly because the dissection of dead bodies was believed to be a sin. The more heavy-handed a dictatorship or the more shortsighted a democracy, Deutsch held, the more justifiable it is for science to insulate itself from society in order to resist oppression.

The need of science, he continued, is to satisfy curiosity, to find out more, to seek the truth. There is also a human need to remain undisturbed in one's own beliefs. Thus, it becomes a question of how much new knowledge should be spread, and how soon. Science can be fed to the public, and the public can eventually become more tolerant of it. Certainly it is worthwhile to lie if the alternative is to be burned at the stake, but it is not worthwhile if the only cost consists of an irate Congressman severing an appropriation, unless it involves an utterly vital matter on which many lives depend. Secrecy defends the weak, but as Kettering said, whenever one locks a laboratory door, one locks more knowledge in than out. Secrecy can have catastrophic cognitive effects. Oppression has cognitive costs. Thus, what is needed is a two-way dialogue—between the scientists, who may know more about certain aspects of what they are doing, and the public, whose members may know more about the consequences of science for the daily life of society.

Scientists who refrain from such a dialogue may lose a great deal more knowledge than they might think.

Before the close of the conference, White asked to express a gratitude that perhaps the conference organizers could not. On behalf of the American and British participants who could not speak German, he wished to apologize that, due to their insular culture, they were unable to speak so important a language as German. He thanked the non-English-speaking members for their courtesy and patience in conducting the entire length of the conference in English.

PART IV

Conclusion

Chapter 19

On Coping with Science as a Task of Policy: A Tentative Summary

Karl W. Deutsch and Andrei S. Markovits

What have we learned in the three days of our conference? Or at least, what do some of us think we learned? What did we seek to find out about so large a problem in so short a time?

We had started from a concern shared by many scientists: will science lose public support and encounter mounting opposition, on a large scale and for a long time, through a general change in the climate of opinion? After all, we are currently witnessing in most advanced industrial societies of the West an opposition to nuclear power plants; resistance to the construction of new airports, large buildings in cities, and some biological laboratories; and opposition to new superhighways and power dams that may endanger a rare species of small fish. This is best exemplified by the much-publicized conflict in the United States over the building of the Tellico Dam at the possible expense of the snail darter. The cumulative impact of all this opposition prompts the question, is a broad change in public attitudes toward science and technology imminent, if indeed not already under way?

Our conference, then, was designed to ask whether there have been changes in the climate of opinion in the past, and what we can learn about their causes and current manifestations which, we believe, already present important policy choices for our future existence.

OUR ORIGINAL IMAGE

We set out with a rather simplified set of images. The Middle Ages, we thought, were characterized by a predominant mood of fear and distrust of innovations in science and technology. The Renaissance and the Age of Enlightenment—the age of Newton, Voltaire, Jefferson, and Lessing—on the contrary, we believed, were epochs seemingly dominated by a general trust in science. Moreover, the nineteenth century, we thought, was deeply divided in its views. The age of the Industrial Revolution in Europe and America had also been the age of Romanticism and its continuing protest against technology and scientific rationality. While James Watt, Robert Fulton, and George Stephenson ushered in the Age of Steam, their contemporary William Blake wrote of "dark, satanic mills," and Edgar Allan Poe created his symbol of mechanized nightmare in his story "The Pit and the Pendulum."

The twentieth century, as we then saw, followed a zigzag course. At first, there had been very large waves of innovations that spread rapidly and were readily accepted. Electric light, the telephone, the automobile, and the airplane are but a few examples from this period of scientific and technological development. Later on, in the 1930s, 1940s, and 1950s, this receptive mood continued. Sulfa drugs, penicillin, and poliomyelitis vaccines were wanted and welcomed, as were sound and color movies, radar, jet airplanes, television, large electronic computers, nuclear energy used for peaceful purposes, and rockets for space navigation.

The 1960s and 1970s, by contrast, were decades of mounting concern. Threats to the natural and historic environment encroached on the countryside and towns. Jet planes were noisy, television was insufficiently artistic and intellectual, and computers threatened individual privacy and freedom. Nuclear power plants were seen as highly unsafe, and landings on the moon as a waste of public resources. Books were written with the intent of showing that physicians caused some diseases and that industry—willfully or by its mere existence—poisoned our water and our air. Some of these warnings proved to be grounded in fact; others combined limited amounts of truth with large measures of exaggeration. The latter increased the books' sales; they seemed to say what people wanted to hear. What, then, were the reasons for these drastic changes in opinion across the decades and centuries? This is how our inquiry started.

A MORE DIFFERENTIATED VIEW

To look for an answer, we brought together a number of experts from various countries and academic disciplines. They did what all good experts do: they made us qualify our questions. Our original picture had been too simple. It had resembled a poster, in a place where an etching with many fine lines and shadings was needed. Trust and fear toward science and technology, they reminded us, had always coexisted, albeit in changing proportions. With respect to the reality of these shifts, however, our original image did survive its critical discussion and scrutiny.

Moreover, it was pointed out that from antiquity until well beyond the middle of the nineteenth century, science and technology had been very far apart. Only in the last hundred years, with the rise of large science-based industries in chemistry, electronics, pharmaceuticals, energy, aeronautics, and other fields, have science and technology come close together and been perceived, by general opinion, as one.

Yet toward this single modern complex of science and technology—which for brevity's sake we shall call "science"—popular attitudes have become more discriminating. Experts on public opinion research from several countries confirmed this finding, which prevails among many nations and apparently even cuts across social systems, East and West.

In general, it seems that there may be something like a *scale effect* in the evaluation of science and its applications. Results that seem to enhance the stature of individuals, increase their options, and affirm a sense of control over their own lives are most often gladly accepted. Here we find the transistor radio, the television set, the electric light, the telephone, and the electronic pocket calculator. Hardly anyone says, "I am afraid of my pocket calculator." Adjustments to scientific or technological change are welcomed if they take place slowly enough to preserve the dignity of individuals, that is, their autonomous control over their own actions, thoughts, and feelings. Similarly, science is favored and even demanded by the public as an aid to the protection of individuals and the environment.

Big technology, on the contrary, is feared and resisted if its huge projects seem to dwarf individuals, make them feel insignificant or helpless, or force them to adjust to exigencies with undignified haste. Science provokes similar resistance, we may surmise, if it increases inequality among human beings, enhances the distance between rich and poor or between elite and non-elite, and thus deepens social alienation. In this context, even talent searches, aptitude or achievement tests, and schools for the gifted may come to be resisted.

MASS EFFECTS AND MASS CONCERNS

Though the public is more discriminating, present-day attitudes toward science have become more powerful in politics. Science, like many other human activities, has become to a considerable degree more large-scale and more capital-intensive. It thus needs more public support in a more sustained manner, both for its operations and for the training of its practitioners. At the same time, however, science and its effects now touch the lives of much larger numbers of people far more directly. These new *mass effects* of science are coming at a time when social mobilization, needs and aspirations, and the capabilities and political awareness of large numbers of people are increasing. The needs of science, its mass effects, and the growth of mass politics are all driving towards a series of potentially constructive or destructive encounters.

THE LIMITED CONTROLLABILITY OF SCIENCE

One symptom of this development is the widespread demand to subject science to closer governmental and popular control. Here conservative spokesmen of preindustrial tradition and radical representatives of new populism seem to agree that science can be controlled, and the only difference pertains to the question of who ought to control it.

This, too, we discussed at our conference: what is decidable in the course of science? It is easier, we agreed, to stifle science than to steer it. However, even stifling science is not easy. In a world of rivalry among several superpowers and middle-level powers, of 150 competing nation-states, of differing ideologies and social orders, and of increasing industrial and economic competition, no major discovery or invention can remain concealed for long. If promising, such a discovery or invention is not likely to remain undeveloped for long. The only option for national policy is to accelerate particular developments, such as that of nuclear energy during and after World War II, or else retard them, as the international agreements on the nonproliferation of nuclear weapons are meant to do. (The virtual cessation of the large-scale use of poison gas after World War I, and the possible broader slowdown in the development of chemical and bacteriological warfare, are rare but perhaps hopeful exceptions.)

THE AUTONOMOUS DYNAMICS OF SCIENCE

All such controls of science by politics or public opinion have thus far been weak and, even so, most often negative. It is very rare indeed that a major positive advance in science can be made on time and on command. The best chance for accelerating such a development,

James Bryant Conant wrote long ago, occurs when the "degree of empiricism" with regard to a problem is low. That is, acceleration is fostered when the basic theoretical aspects of the matter are already clearly understood, and only the practical "know-how" is still lacking for its large-scale development and application. These conditions existed in 1940–1941, according to Conant, with regard to nuclear energy, but not with regard to the synthesis of penicillin. Hence the development of nuclear energy could be accelerated, as indeed it was, but penicillin synthesis had to wait for the elucidation of its chemical structure, which came only some years later.[1]

The *autonomous dynamics* of science had already appeared in these early examples from Conant's analysis. The participants of our conference seemed to agree that these autonomous dynamics are usually more important, and have weightier effects on the course of scientific development, than any efforts at positive control or guidance from outside. What is desirable, they pointed out, is not always quickly feasible. Neither a cure for cancer nor a cheap, efficient, and lightweight solar power accumulator has yet been discovered, though they have been sought for almost a century or more. Society has not divined the regularities of nature that may underly these failures or delays.

More generally, it was said, the *pathways* toward a new discovery or invention cannot be known wholly in advance. The development of science has more in common with *theory of search* into the unknown than with any deterministic theory of planning.

SOME CHOICES THAT WE HAVE

What we *can* decide remains substantial: whether to seek or not, on what scale, by what means, and at what time. We make these decisions often, and they may have larger consequences than we believe.

All such thoughts are fraught with risks of error. We can only choose on which side we wish to err. In more technical terms, we must often choose between the risks and costs of an error of the first or of the second kind: that is, the risks and costs of accepting something as true which is false, versus the risks and costs of rejecting something as false which is in fact true. Should we accept as beneficial a particular type of medical therapy or engineering project, or should we reject them as useless or dangerous? Either choice runs the risk of being wrong. Which risk we choose depends on its expected costs and those of its alternatives.

Let us note that doing nothing is also a choice, one which is apt to entail costs and perhaps larger consequences. If it is good to require "environmental impact statements" or other forms of technology assessment before allowing some major construction project to proceed, would it not also be reasonable at the same time to stipulate an "inactivity impact statement" or "inactivity assessment"? Not putting a new roof on a house before a hard winter may prove more costly than installing one would have been.

Do we, however, know enough to choose? The large systems of technology already built or under construction, such as extensive computer programs or nuclear power plants, are so complex, it was said, that no one can understand what is actually happening within the interplay of tens of thousands of smaller elements within them—not even their architects. The very complexity of these large manmade systems may make their function *opaque*, at least to human beings; this point was made vigorously in the paper by Joseph Weizenbaum. There may be a growing mismatch, he seemed to imply, between the increasing complexity of human artifacts and the limited capacities of human minds. His views were strongly contested, but they were more than a mere warning against trusting "big science" and its works. In their deeper implications, Weizenbaum's points represented a profound intellectual and existential challenge to scientists, including Weizenbaum himself, to develop a *theory of the transparency of large systems*—the conditions favoring or opposing them, their limits within each situation, and our ability to extend them.[2]

WHO SHOULD CHOOSE—AND HOW?

Though choices concerning science are limited, they must be made by some persons and institutions, as well as through some methods. Since the inner dynamics of science are important, there seemed to be a case for leaving most of these choices to the *autonomous dynamics* of science and having the rest made from within, by the scientists themselves.

Yet scientists themselves also constitute an interest group in their relations with other social and economic actors. They compete for money and other limited resources with activities toward other ends. Within science, there are sometimes schools of thought and various organizations, all with their special biases. Scientists with substantial reputations, or those leading major research projects or institutions, also form a kind of *establishment*, it was stated, with observable advantages that even gifted and diligent outsiders rarely share. Different scientists gave different weights to each of these matters, but there was general agreement that such concerns were not wholly unrealistic.

It was also pointed out, however, that ability, motivation, and effort are unevenly distributed. At least in some fields, approximately 50 percent of the important contributions are made by only about 10 percent of the scientists. These highly productive scientists are making contributions not only because they are well connected and financially supported. More often, they are doing so because they have interesting ideas and have connections among themselves as a direct or indirect result of mutual research interests and intellectual affinities. The growth of science, in the main, does not occur through inbreeding or favoritism. There were some skeptics but no cynics in this matter at the conference.

The need for outstanding scientific performance seems likely to continue despite its uneven distribution, and its effects are reinforced by a need for a somewhat uneven allocation of material support. A nearly equal assignment of resources among all scientists might have one of two results.

On the one hand, it might force each of them to limit his or her work to questions which could be answered within a reasonable time by the labor of one person or a very small number of people; they would hold down the production of knowledge to a handicraft scale. Alternatively, for dealing with broader questions, scientists might have to pool their resources into some collective or group, with those most skilled in handling group meetings emerging as the leaders. There are few, if any, cases in which such egalitarian collectives have produced major contributions to the solution of problems requiring research on a large scale. More common has been the practice of entrusting major scientific tasks to a few principal investigators in charge of an institution or project, to which they then recruit other colleagues, granting to them larger or smaller shares in decisionmaking, often in accordance with experience and professional standing. This method poses a risk that hierarchy and bureaucratization may impede communication among scientists. Nevertheless, large scientific projects in the East and the West have most frequently been carried out in this manner, often with success.

Even so, the protests against "elitism" and "big science" by some "little" scientists, as well as by some dissenting scientists with prominent reputations, have been part of the democratic process, making it unlikely that the trees of any establishment or group of insiders will grow into the sky. What emerged from the discussion was the clear impression that control over the attention, resources, and priorities of science today cannot be left entirely to the scientists alone, just as it was not left to their exclusive discretion in most past epochs.

What other mechanisms of guidance or control are there to supplement the partial but indispensable autonomy of science and of scientists? The commercial *free market* might seem to be such a mechanism, steered in its turn by the preferences of consumers but also by the uneven distribution of ownership of wealth, political power, and the existence of oligopolistic organization in all advanced industrial societies. No one thought that market mechanisms could be entirely dispensed with, but neither did anyone seem ready to trust them blindly or completely. We did not believe that the market would lead science very far, but we also knew that all existing social and economic *planning mechanisms* have their own large weaknesses. Rather, it seemed a question of searching for a *combination* of several methods which one might hope would jointly work better. No such combination thus far has been found, but the search is apt to continue.

There remains the *political process*. Politics and politicians must choose time and again among a market friendly to well-to-do consumers, the autonomy of science, the claims of the poorer strata and countries, and perhaps the large conceptions of economic or military planners. (Such plans and planners exist, of course, also in the private sector of many countries and in the large multinational corporations.)

If politicians must find solutions and gain their acceptance, how and where can they obtain the necessary knowledge to do so? They may know enough about voters, but how much do legislators and their constituents know about science? Here public opinion expert Max Kaase reported an interesting finding. The better educated the respondents to survey questions are, the more likely they are to say that they expect many future benefits from science despite specific worries about particular problems. They also say—and increasingly so with their level of education—that they do not feel competent to make decisions about science.

Recently, a Europe-wide survey, including countries from both the East and the West, showed the existence of about 43 percent of such "Socratics" with regard to this topic—that is, persons who admitted that they knew nothing about the complexities of science, or certainly not enough to give intelligent answers, let alone make decisions. Socrates himself in his day had a smaller proportion of his compatriots on his side. Here something about democracy has changed, according to the survey data.

However, while the knowledge that one does not know enough about some important problem may be the first step toward seeking and accepting more knowledge, our problems themselves have grown vastly in volume and in kind.

A NEW SCALE OF PROBLEMS:
ARE WE IN THE MIDDLE OF AN
EVOLUTIONARY JUMP?

An especially challenging view of the growth of our problems— and of our capabilities—was presented by the mathematical biophysicist

John Platt. He listed a number of basic functions or capacities of living organisms as they had arisen in the course of evolution, and then examined the recent major changes mankind had undergone with respect to them. His functions were startling.

One such function was the dimension of the *speed* with which organisms were able to move through space. This speed had increased slowly from the drifting of the early amoebae in the primeval waters to the swimming of fish and the crawling of reptiles, on to the 60 kilometers per hour of swift horses and other fast land animals, the more than 100 kilometers per hour of some birds in flight, the 150 kilometers per hour of the fastest nineteenth century railroad engines, and the 180 kilometers per hour of commercial aircraft in the 1920s. Since then, in roughly sixty years, human speed has increased by a factor of about ten, through such commercial aircraft as the Concorde, and by a factor of about one hundred, through spaceships whose 18,000 kilometers per hour exceeds the velocity needed to escape the gravitational pull of our planet.

Another dimension is the range of *habitats*, that is, of environments which a life form can inhabit. The first biological "invention" here was the cell with walls protecting its contents. Some microorganisms can live in a wide range of environments, and later, mammals and birds developed with warm-blooded bodies that could maintain a constant temperature. Generally, however, the larger and more complex the organism is, the smaller the range of environments in which it can survive. An exception to this rule is the human race, which has thrived for approximately one hundred thousand years with the help of clothing and manmade shelters, even in arctic regions and in sunbaked deserts. But in the last sixty to eighty years, people have put part of their mass transportation into the air, some of their mining on the bottom of the sea, and, during only the last twenty years, some of their voyages into outer space and on the surface of the moon. Protected by new types of vessels and life-support systems, man has in a remarkably short and recent time radically expanded his range of actual and prospective habitats.

New ranges of speed and habitat were made possible through new sources of energy. Higher energy concentrations via metabolism were achieved successively by plants, herbivores, and carnivores. The use of fire was discovered by man's ancestors perhaps five hundred thousand years ago or earlier. The use of animal power, water power, wind, and charcoal came perhaps within the last five thousand or ten thousand years, gunpowder within the last eight hundred, mineral coal and steam within the last three hundred, and electric power within the last one hundred years. The large-scale use of petroleum and natural gas is only about seventy years old, and nuclear energy less than forty. The explosive power of a bomb increased by a factor of five thousand—from the ten tons of TNT in a World War II blockbuster to the one hundred megatons in a thermonuclear weapon reportedly tested in the early 1960s.

This last example reminds us how closely linked is the increase in human capabilities to the growth of dangers. But already a vast new source of energy is being tapped—the sun—which may prove practically inexhaustible. In terms of biology and history, the sweep and speed of all these recent developments in energy seem unmistakable.

If energy developments are giving us a partial equivalent to new muscles, a change in our capacity to perceive events at a distance has given us something like additional sets of eyes. Some relatively primitive organisms developed spots sensitive to light several millions of years ago. Specialized cells and eventually eyes followed in the course of the evolutions of several quite different life forms—insects, fish, amphibians, reptiles, birds, and mammals. Human beings have developed telescopes with their modern sophistication only within the last fifty years. However, during the last eighty years, we have learned to use the entire electromagnetic spectrum from electron microscopes to radio astronomy to gather information. In this respect, never before has there been so great a change in so short a time.

Through polymers, synthetics, and plastics, chemistry has given us a new series of fibers and materials in the last fifty years, comparable to man's original acquisition of metals and fiber plants thousands of years ago. New antibiotics have changed parts of medicine, and new mind-altering drugs have promoted new possibilities—as well as abuses—of psychotherapy.

Among the slowest changes in biological evolution have been those relating to heredity, since the "invention" of sexual reproduction. Within the last thirty years, new possibilities for changing the genetic code, and thus the hereditary characteristics of life forms, have been created through DNA manipulation. New strains of microorganisms already created by this method produce interferon and insulin, two medically important substances. Together with the new techniques of cloning, these methods are beginning to transform our knowledge of living organisms and our dealings with them.

Perhaps the most crucial change has occurred in our ability to deal with *information*—to transmit, store, and process it. Within the last forty years, large computers have been developed, and within the last twenty years, transistors, miniaturization, and microchips have enabled us to make computers small while preserving and even enhancing their already large capacities. In terms of evolution, this is the first major change since the development of nerves and brains. In terms of human history, it is comparable to the development of language and writing. It is as if humanity had acquired some additional layers of brain tissue, or a new piece of brain, albeit one-sided and limited in its functions.

All these changes have transpired within the last few decades. They have come with unprecedented speed, and they have all converged within more or less the same short time. Many of their full effects are yet to come, and again, they will all arrive more or less at the same time. There has been no such coincidence of so many and such large changes in the previous five thousand years of the recorded history of mankind and, as far as we know it, in the 4 billion years of the evolution of life on this planet.

Our problem today, John Platt concluded, is not to choose whether we should enter a stream of uniquely rapid and simultaneous changes in science and technology, with their effects on our economies, societies, and cultures. The problem is rather to realize that we are already in the midst of this torrent of change, perhaps in the midst of a historic mutation or an evolutionary jump. Our task now is to learn what to do about this substantial qualitative change presently shaping our future.

It is a task that will require from us new capabilities for orientation to this new situation and for coping with its challenges. Our lag in developing these capabilities could be a major source of danger. If we are to develop them, more, not less, science and technology will be among their indispensable sources.

Platt's assessment of the situation met with agreement as well as with lively dissent. Other periods of accelerated change in human history and culture were recalled, such as those of Athens in 600–300 B.C., Florence in A.D. 1200–1600, and perhaps Britain and France from A.D. 1500 to 1900. The changes in science and technology during these epochs were smaller than those in the twentieth century, and they arrived three to five times more slowly. Nevertheless, something about our own time might yet be learned from comparative *cultural acceleration studies* of such earlier periods of rapid change, perhaps carried out collaboratively by historians and scientists.[3]

Very rapid change, it seems, may favor extremes in human behavior, both in fear and in daring. When a car travels quickly along a road, some passengers may hold on tightly to the safety straps and even protest to the driver, while others may want to go even faster. Both protests and urgings, then, would not indicate that the car was moving slowly, but rather that it was going quickly and perhaps accelerating. Here another piece of public opinion data, reported by Max Kaase, appeared relevant. Younger respondents, such as those between twenty and thirty years of age, seemed to notice fewer ongoing changes than did their elders. To what extent they were less observant, had fewer memories as a basis for comparison, or simply took a higher speed of change for granted was not clear, but might merit further study.

In any case, major decisions about the support of science and technology, and to a limited extent about the directions of their development, will have to be made by the people now living, beset by contradictory pressures and by uncertainty.

QUESTION OF VALUES

Decisions reached under uncertain conditions force us to reveal our values. On which side of a question do we prefer to take our chances? If we must err, on which side would we prefer to do so?

A distinguished scientist argued that his great love for driving sports cars was in and of itself an esoteric hobby at best, but one which would become a moral outrage, possibly even a criminal act, if it were to lead him to entrust such a potentially lethal machine to a fourteen-year-old child. Using such an analogy, this participant saw compelling parallels between this example and the public's potential recklessness in its handling of science and technology's applications. The analogy also connoted the public's decreasing control over science and technology. Others replied in effect that the voters and governments of the German Federal Republic, France, the United Kingdom, the United States, and other industrial democracies could not be equated with a fourteen-year-old child in any way. The question then arises as to how prudent, realistic, open-minded, and capable of agreement we consider our contemporaries to be. Some felt that we must be somewhat optimistic about our fellow human beings, while others thought that the consequences of such trust would be catastrophic.

Here we came to a critical point in our deliberations. The question was not whether people should trust science, for we agreed that blind trust was dangerous. Rather it was whether people could be trusted with science. It is one form of the more general question posed long ago by Dostoevsky's Grand Inquisitor.[4] Our answer must hinge on our conception of what it means to be human. On our answer depends not only what we do, but also what we shall become, each one of us and all of us together. Citizens, legislators, and political leaders will make most of these decisions, both day to day and in the crises to come.

NOTES

1. James Bryant Conant, *Science and Common Sense* (Cambridge, Mass.: Harvard University Press, 1951).

2. Some steps toward such a theory exist in the work of Herbert Simon and other theorists of systems or communication. See, for example, Herbert Simon, *The Science of the Artificial* (Cambridge, Mass.: MIT Press, 1969); Karl W. Deutsch and Bruno Fritsch, *Zur Theorie der Vereinfachung* (Meisenheim am Glan: Verlag Anton Hain, 1980.) Further developments in this direction should be among the significant tasks for research.

3. Attention to the effects of accelerated cultural change has already been paid, though in different contexts, in such works as Alvin Toffler, *Future Shock* (New York: Random House, 1970).

4. Fyodor Dostoevsky, *The Brothers Karamazov,* translated from the Russian by Constance Garnett (New York: Modern Library, 1937).

Index

About the Editors and
Participants

Andrei S. Markovits is Assistant Professor in the Department of Government at Wesleyan University and Research Associate at the Center for European Studies of Harvard University. He obtained B.A., M.B.A., M.A., M.Phil., and Ph.D. degrees from Columbia University. He has recently written numerous articles on comparative education, student politics, international labor problems, and Jewish affairs. The coeditor of *Problems of World Modeling: Political and Social Implications*, Professor Markovits is currently conducting a cross-national study of the reactions to the television series "Holocaust" and writing a book on the post–World War II German labor movement.

Karl W. Deutsch is the Stanfield Professor of International Peace in the Department of Government at Harvard University, and the Director of the International Institute for Comparative Social Research at the Science Center Berlin. He holds a Jur.Dr. from Charles University in Prague, a Ph.D. from Harvard University, an Ll. D. from the University of Illinois, and several honorary degrees. Professor Deutsch's publications include *The Nerves of Government*, *The Analysis of International Relations*, *Politics and Government*, and *Tides Among Nations*. His most recent research involves global modeling, international peace, and the politics of emerging nations.

Harvey Brooks is the Benjamin Peirce Professor of Technology and Public Policy and Professor of Applied Physics at Harvard University. He received an A.B. from Yale University and a Ph.D. in physics from Harvard University. Professor Brooks is currently researching energy policy, science, technology and public policy, science policy, and nuclear energy and its future.

I. Bernard Cohen is the Victor S. Thomas Professor of the History of Science at Harvard University and a faculty member of the Kennedy School of Government of Harvard University. He received an S.B. in mathematics and a Ph.D. in history of science from Harvard University. Professor Cohen is researching the history of scientific ideas, the development of science in America, the Scientific Revolution, revolutions in science, the scientific work of Isaac Newton, historical aspects of science and public policy, and the relations between the sciences and the social sciences.

James A. Cooney is Assistant Director of the Aspen Institute for Humanistic Studies in Berlin. He obtained a B.A. in government from Harvard University, studied at the Free University of Berlin as a Fulbright Scholar, and received an Ed.M. in education and social policy from Harvard University. Mr. Cooney is presently a candidate for the Ph.D. at the Massachusetts Institute of Technology. He is researching business–state relations and nuclear energy policy in West Germany.

Werner Ellerkmann is Site Director of the Joint Research Centre–Ispra Establishment (Commission of the European Communities) in Ispra (Varese), Italy. He holds a Dr. Jur. from Bonn University and has also studied law at the University of Paris, where he passed the first and second state examinations. Dr. Ellerkmann's studies and writings have focused on civil and criminal law, international law, and research administration.

Wolfram Fischer is Professor of Economic and Social History at the Free University of Berlin. He studied at the University of Heidelberg and received his Dr. Phil. from the University of Tübingen. He also studied at Göttingen University and the London School of Economics and received a Dr. rer. pol. from the Free University of Berlin. Professor Fischer's latest publications include *Economic History of Europe 1950–Present*, which he edited. He is the author and editor of *World Economic History of the 20th Century*.

Bruno Fritsch is Professor of Economics at the Swiss Federal Institute of Technology (ETH) in Zurich and president of the Swiss Association for Future Research, also in Zurich. He studied economics, philosophy, and mathematics and was awarded a Dr. rer. pol. from the University of Basel. Professor Fritsch is presently teaching and researching international economics, energy economics, economic development, interactions between economic activities and ecological systems, world modeling, and the relation between energy, resources, and knowledge. His most recent publications include *Growth Limitation and Political Power* and various articles on social costs, planning, simulation, ecology, growth, and research on the future.

Peter Glotz, who received his Dr. Phil. from the University of Munich, is Senator for Science and Research in West Berlin, a position he has held since May 1977. Between 1972 and 1977 he was deputy chairman of the Bavarian Social Democratic Party (SPD) and a member of the Bundestag in Bonn. Between 1974 and 1977, he was Parliamentary State Secretary in the Federal Ministry of Education and Science. Among Dr. Glotz's publications are numerous articles and three books. His activities and interests in public life have been concentrated in the fields of youth, education, and the state's role in setting an optimal climate for teaching and research.

Elisabeth Helander is Research Director at the Academy of Finland. She studied at Helsinki University, where she earned a Ph.D. in nuclear chemistry and physics. Her current research interests include science policy studies and environmental research.

Max Kaase is Professor of Political Science at the University of Mannheim, and the Executive Director of ZUMA, the Zentrum für Umfragen, Methoden und Analysen (Center for Survey Research, Methods and Analysis). He obtained a Dr. rer. pol. from the University of Cologne and finished his "habilitation" at the University of Mannheim. Professor Kaase is currently teaching political sociology, with emphasis on electoral sociology. His research involves political participation, mass communication, democratic theory, and social science research methodology. He is the author of various books and numerous articles, of which the latest in English is "The Crisis of Authority in Western Liberal Democracies: Myth and Reality," in Richard Rose, ed., *The Burdens of Governing* (forthcoming).

Wilhelm A. Kewenig is Professor of Law at Kiel University and the Director of the Institute of International Law. He studied law and political science at the universities of Bonn, Freiburg, Cologne, Paris, and Beirut and also attended the Harvard Law School. Professor Kewenig's current interests include the new economic world order, sea law, United Nations law, human rights, the law of mass communications, and parliamentary law.

Wolfgang Krohn is lecturer in Science in Technology of the Renaissance at the University of Munich and a member of the staff of the Max Planck Institute for Research on the Survival Conditions of the Scientific–Technological World. He studied at the universities of Göttingen, Marburg, and Hamburg and obtained a Dr. phil. from the University of Hamburg. Dr. Krohn's present research involves the origins and structure of early modern science, relations between humanism and artist-engineers in the Renaissance, the legitimation crisis of science and technology, and relations between scientific and social progress.

Karl-Hans Laermann is Professor of Statics of Structures at the University of Wuppertal and a member of the German Bundestag, representing the Free Democratic Party (FDP); he also serves as the Vice Chairman of the Bundestag Committee on Research and Technology. Professor Laermann is presently researching experimental stress analysis, statics of structures, nonlinear elastics, viscoelastic response of materials, and large deformations.

David S. Landes is the Robert Walton Goelet Professor of French History and a Professor of Economics at Harvard University. He obtained an A.B. from the City College of New York and an A.M. and a Ph.D. from Harvard University in history. Professor Landes is currently doing research in population, food supply, and natural resources in historical perspective, in cultural determinants of entrepreneurship, and in the history of the Arab–Israeli conflict. He is also presently working on a textbook on Western civilization.

Peter Laslett is Reader in Politics and the History of Social Studies at Cambridge University, a Fellow of Trinity College, and a Fellow of the British Academy. He is also the Director of the Cambridge Group for the History of Population and Social Structure. He received an M.A. in history from Cambridge University.

Gerhard Leminsky, who received a Dr. rer. pol. from the University of Hamburg, is a member of the Institute for Economic and Social Research of the German Trade Union Federation and the editor of the Trade Union monthly *Gewerkschaftliche Monatshefte*. He also serves on trade union committees, and on government commissions and supervisory boards as a trade union representative. Dr. Leminsky's research centers on the role of unions in industrial relations systems, workers' participation and collective bargaining, employment policy and humanization of work, science policy, and the development of technology and unions.

Hans Wolfgang Levi is a professor at the Technical University of Berlin and Director of the Hahn Meitner Institute for Atomic Research. He received a Dr. Ing. in chemistry from the Technical University of Berlin, where he also studied nuclear chemistry and nuclear chemical engineering. Professor Levi is currently writing a textbook about nuclear chemical engineering and researching problems related to nuclear safety.

Robert Lopez is the Sterling Professor of History at Yale University. He obtained a D.Litt. from the University of Milan and a Ph.D. from the University of Wisconsin. He has also studied literature, art, and law. Professor Lopez is the author of thirteen books and more than a hundred articles.

Claus Offe is Professor of Political Science and Sociology at the University of Bielefeld. He studied at the Free University of Berlin, received a Dr. phil. from the University of Frankfurt, and finished his "habilitation" at the University of Konstanz. He has also done research at the Max Planck Institute at Starnberg. Professor Offe is presently researching policy studies in the areas of labor market problems, regulation of industrial relations, comparative education, and social policy. He is also working on theories of "postindustrial" social formations and is interested in general political sociology.

Horst Ohnsorge is Director of the Research Center of Standard Electric Lorenz AG in Stuttgart. He has also served as Research Director for Communications systems at AEG-Telefunken in Ulm and as Director of the Heinrich Hertz Institute for Communications Technology in Berlin. Dr. Ohnsorge studied electrical engineering in Giessen and at the Technical University at Darmstadt, where he received his Dipl. Ing. and Doctoral degrees. The contributor of many articles to professional journals, Dr. Ohnsorge has held numerous research positions in universities and industry and is the recipient of sixty-six patents. His major research interests focus on problems of coding, information theory, and information systems, especially optical channels and image transmissions.

John Platt is Lecturer of Physiology and Socio-Medical Sciences at the Boston University Medical School. He obtained a B.S. and M.S. in physics from Northwestern University and a Ph.D. in physics from the University of Michigan. Dr. Platt's primary research involves problems concerning the future, science and society, and rates of change.

Ulrich Steger is a member of the Social Democratic Party (SPD) and a representative to the German Bundestag. He is also the chairman of the SPD Research and Technology Committee. He obtained a Dr. rer. oec. from the Ruhr University in Bochum, where he studied theories of political economy and questions of structural politics. Dr. Steger is presently studying science and technology, especially as they relate to energy problems.

Jean Stoetzel is Distinguished Professor at the Sorbonne in Paris, a member of the French Academy of Moral and Political Sciences, and Vice President of the French Commission to UNESCO. He studied philosophy and obtained his doctoral degree from the Sorbonne, where he continues to teach sociology and social psychology. Professor Stoetzel is currently involved in researching various aspects of public opinion and the distribution of income.

Shepard Stone is Director of the Aspen Institute for Humanistic Studies in Berlin. He received an A.B. in American history from Dartmouth College, a Ph.D. in European history from the University of Berlin, and honorary doctorates from the Free University of Berlin and the University of Basel. He was named an Honorary Professor by the city of Berlin in April 1978.

Stephen Strickland is Vice President of the Aspen Institute for Humanistic Studies in Washington, D.C. He holds a B.A. in political science from Emory University and an A.M. and Ph.D. in political science from Johns Hopkins University. Dr. Strickland's interests lie in the area of science policy and policymaking.

Klaus Traube is a physicist and freelance writer. He obtained a Dipl. Ing. from the Braunschweig Technical Institute, where he studied mechanical engineering, and a Dr. Ing. from the Munich Technical Institute. Dr. Traube is currently researching aspects of the interaction between technology and society.

Thomas A. Trautner is a professor at the Free University of Berlin and Director of the Max Planck Institute for Molecular Genetics. He was awarded a Dr. rer. nat. from the University of Göttingen and finished his "habilitation" at the University of Cologne. He also studied at the University of Illinois, Stanford University, and the University of California at Berkeley. His major subjects of interest have been biology, microbiology, biochemistry, and genetics. Most of Professor Trautner's research is in the field of molecular biology and gene technology.

Peter Weingart is Professor of Science and Science Policy at the University of Bielefeld. He studied economics, sociology, business administration, and constitutional law at the University of Freiburg and the Free University of Berlin, as well as at Princeton University, where he was a University Fellow. He received an M.A. and a Dr. rer. pol. from the Free University of Berlin. Professor Weingart is presently researching topics in the sociology of science, political sociology, science and education policies, and the philosophy of science. He is the author of numerous books and articles and coeditor of the *International Yearbook of the Sociology of Sciences*.

Joseph Weizenbaum is Professor of Computer Sciences at the Massachusetts Institute of Technology. After obtaining his B.S. and M.S. degrees from Wayne University, Professor Weizenbaum spent eight years as a systems engineer at the Computer Development Laboratory of the General Electric Company. His numerous research interests and topics of writing have included artificial intelligence, natural-language understanding by computers, the structure of computer languages, and the social implications of computers and cybernetics. He is the author of *Computer Power and Human Reason*, and the composer of SLIP, a list-processing language, and ELIZA, a natural-language processing system.

Harrison C. White is Professor of Sociology at Harvard University and a member of the National Academy of Sciences and the American Academy of Arts and Sciences. He obtained a B.Sc. in physics and a Ph.D. in theoretical physics from the Massachusetts Institute of Technology, and a Ph.D. in sociology from Princeton University. Professor White is currently the Director of a project in Cross-Disciplinary

Training to Model Complex Systems at Harvard University and the Director of the National Research Council's Assembly of Behavioral and Social Sciences. He has been a member of the editorial board of the *Journal of Mathematical Sociology* since 1970.

About the Science Center
Berlin

The Science Center Berlin (Wissenschaftszentrum Berlin), a nonprofit corporation, serves as a parent institution for institutes conducting social science research in areas of significant social concern.

The following institutes are currently operating within the Science Center Berlin:

1. The International Institute of Management,
2. The International Institute for Environment and Society,
3. The International Institute for Comparative Social Research.

They share the following structural elements: a multinational professional and supporting staff, multidisciplinary project teams, a focus on international comparative studies, a policy orientation in the selection of research topics and the diffusion of results.